Algebra 2

LARSON
BOSWELL
KANOLD
STIFF

Applications • Equations • Graphs

Chapter 9 Resource Book

The Resource Book contains the wide variety of blackline masters available for Chapter 9. The blacklines are organized by lesson. Included are support materials for the teacher as well as practice, activities, applications, and assessment resources.

McDougal Littell
A HOUGHTON MIFFLIN COMPANY
Evanston, Illinois • Boston • Dallas

Contributing Authors

The authors wish to thank the following individuals for their contributions to the Chapter 9 Resource Book.

Rose Elaine Carbone
José Castro
John Graham
Fr. Chris M. Hamlett
Edward H. Kuhar
Ann C. Nagosky
Karen Ostaffe
Leslie Palmer
Ann Larson Quinn, Ph. D.
Chris Thibaudeau

ISBN: 0-618-02017-9

89-PBO- 04

Contents

9 *Rational Equations and Functions*

Contents

Contents

Descriptions of Resources

This Chapter Resource Book is organized by lessons within the chapter in order to make your planning easier. The following materials are provided:

Tips for New Teachers These teaching notes provide both new and experienced teachers with useful teaching tips for each lesson, including tips about common errors and inclusion.

Parent Guide for Student Success This guide helps parents contribute to student success by providing an overview of the chapter along with questions and activities for parents and students to work on together.

Prerequisite Skills Review Worked-out examples are provided to review the prerequisite skills highlighted on the Study Guide page at the beginning of the chapter. Additional practice is included with each worked-out example.

Strategies for Reading Mathematics The first page teaches reading strategies to be applied to the current chapter and to later chapters. The second page is a visual glossary of key vocabulary.

Lesson Plans and Lesson Plans for Block Scheduling This planning template helps teachers select the materials they will use to teach each lesson from among the variety of materials available for the lesson. The block-scheduling version provides additional information about pacing.

Warm-Up Exercises and Daily Homework Quiz The warm-ups cover prerequisite skills that help prepare students for a given lesson. The quiz assesses students on the content of the previous lesson. (Transparencies also available)

Activity Support Masters These blackline masters make it easier for students to record their work on selected activities in the Student Edition.

Alternative Lesson Openers An engaging alternative for starting each lesson is provided from among these four types: ***Application, Activity, Graphing Calculator,*** or ***Visual Approach.*** (Color transparencies also available)

Graphing Calculator Activities with Keystrokes Keystrokes for four models of calculators are provided for each Technology Activity in the Student Edition, along with alternative Graphing Calculator Activities to begin selected lessons.

Practice A, B, and C These exercises offer additional practice for the material in each lesson, including application problems. There are three levels of practice for each lesson: A (basic), B (average), and C (advanced).

Contents

Reteaching with Practice These two pages provide additional instruction, worked-out examples, and practice exercises covering the key concepts and vocabulary in each lesson.

Quick Catch-Up for Absent Students This handy form makes it easy for teachers to let students who have been absent know what to do for homework and which activities or examples were covered in class.

Cooperative Learning Activities These enrichment activities apply the math taught in the lesson in an interesting way that lends itself to group work.

Interdisciplinary Applications/Real-Life Applications Students apply the mathematics covered in each lesson to solve an interesting interdisciplinary or real-life problem.

Math and History Applications This worksheet expands upon the Math and History feature in the Student Edition.

Challenge: Skills and Applications Teachers can use these exercises to enrich or extend each lesson.

Quizzes The quizzes can be used to assess student progress on two or three lessons.

Chapter Review Games and Activities This worksheet offers fun practice at the end of the chapter and provides an alternative way to review the chapter content in preparation for the Chapter Test.

Chapter Tests A, B, and C These are tests that cover the most important skills taught in the chapter. There are three levels of test: A (basic), B (average), and C (advanced).

SAT/ACT Chapter Test This test also covers the most important skills taught in the chapter, but questions are in multiple-choice and quantitative-comparison format. (See *Alternative Assessment* for multi-step problems.)

Alternative Assessment with Rubrics and Math Journal A journal exercise has students write about the mathematics in the chapter. A multi-step problem has students apply a variety of skills from the chapter and explain their reasoning. Solutions and a 4-point rubric are included.

Project with Rubric The project allows students to delve more deeply into a problem that applies the mathematics of the chapter. Teacher's notes and a 4-point rubric are included.

Cumulative Review These practice pages help students maintain skills from the current chapter and preceding chapters.

Tips for New Teachers

For use with Chapter 9

Lesson 9.1

INCLUSION Students may have difficulty writing variation equations from word problems. Practice reading through problems and identifying key elements. Have students underline these elements and define a variable for each. Then reread the problem and at the same time set up the equation with the defined variables.

COMMON ERROR Students often fail to understand that k, the *constant of variation*, is a constant. Review the definitions of constant and variable. Stress the fact that k will never change for a given problem, despite the fact that the variables do.

TEACHING TIP Inverse variation problems can be solved without finding the constant of variation, k. To do so, write the same inverse variation equation for two different sets of variables:

$$y_1 = \frac{k}{x_1} \text{ and } y_2 = \frac{k}{x_2}.$$

If we now solve for k in each equation we have $k = x_1 \cdot y_1 = x_2 \cdot y_2$. Using the last part of this equation we can find a missing value without previously solving for k. In a similar manner, direct variation problems can be solved by using the fact that

$$k = \frac{y_1}{x_1} = \frac{y_2}{x_2}.$$

Lesson 9.2

COMMON ERROR Many students struggle to find the domain and range of rational functions. Tell students to start by finding the horizontal and vertical asymptotes of the graph of the function. For the type of rational functions covered in this lesson, those are the values *excluded* from the domain and range, respectively.

INCLUSION Students may have a difficult time creating an algebraic model for a word problem such as Example 3 on page 542. Ask students to *actively* read the problem, underlining any relevant information. Then, have them explain what is happening in their own words.

Lesson 9.3

TEACHING TIP To help students graph rational functions, break down the process into three steps. First, students should find the x-intercepts of the function. Second, they must find the vertical asymptotes of the graph. Third, they must compare the degrees of the numerator and denominator to determine the graph's end behavior. Number these steps, and follow them consistently when graphing rational functions.

TEACHING TIP Show your students an example of a function with a *point of discontinuity*, such as

$$f(x) = \frac{x^2 - 9}{x - 3}.$$

For this function, when $x = 3$ we have

$$f(3) = \frac{0}{0},$$

which is undefined. The function can also be written as

$$f(x) = \frac{(x - 3)(x + 3)}{x - 3}.$$

Therefore, its graph is the line $y = x + 3$, with a discontinuity—a "break"—at $x = 3$. The function has neither vertical nor horizontal asymptotes.

COMMON ERROR Some students mistakenly believe that the asymptotes of a function are part of the graph of the function. The activity on page 546 will show students that an asymptote is an aide to graph the function, but it is not part of the function.

Lesson 9.4

COMMON ERROR Students often incorrectly cancel like terms from the numerator and denominator without first factoring these expressions. For example, they might evaluate $\frac{x^2 + 4x}{x^2}$ as $4x$.

Substitute a value for x to show students that this is incorrect. Remind students that before they cancel out any terms, they must factor both the numerator and the denominator, if possible.

Tips for New Teachers

For use with Chapter 9

Lesson 9.5

COMMON ERROR When adding or subtracting rational expressions with unlike denominators, some students fail to change the numerator after finding the common denominator. For example, they might write

$$\frac{5}{x+3} + \frac{x}{x-2} = \frac{5+x}{(x+3)(x-2)}.$$

You can show students their mistake by asking them to substitute a value for the variable and evaluate both expressions. Emphasize that whatever factor is used to multiply the denominator must also multiply the numerator.

TEACHING TIP When learning to simplify complex fractions, it might help students to take an extra step to avoid mistakes. Once students have written both numerator and denominator as a single fraction, have them write down the division *horizontally* rather than *vertically*. For instance,

$$\frac{\frac{x+1}{3}}{\frac{x}{9}} = \frac{x+1}{3} \div \frac{x}{9}.$$

To do this division, students must now multiply by the reciprocal of the second fraction. This extra step can be skipped as students become more proficient.

INCLUSION Having to choose between two different techniques to simplify complex fractions—see Examples 5 and 6 on page 564—might confuse some students. Show students both techniques but emphasize that they can always use the first one to simplify a complex fraction. Although the problem might take a little longer, some students would prefer to always use the same method.

Lesson 9.6

TEACHING TIP Here is another method of solving rational equations without having to multiply everything by the LCD. Find the LCD and then set up an equation with just the numerators of equivalent rational expressions if these were to have the LCD as the denominator. For example, if we have

$$\frac{2x}{x+3} - \frac{x}{x+7} = \frac{x^2-1}{x^2+10x+21},$$

then the LCD is $(x + 3)(x + 7)$. If that was the denominator, the numerators of these 3 rational expressions would make up the equation $2x(x + 7) - x(x + 3) = x^2 - 1$. This equation is true because if two rational expressions are equal and have the same denominator, the numerators must be equal.

COMMON ERROR Students will mistakenly claim that there is no solution or that the solution is extraneous if the value of the solution happens to be a complex number. Make sure to complete an example in class where the answer is not a real number. Then discuss the validity of the answer for an exercise or in the context of a real-life situation.

Outside Resources

BOOKS/PERIODICALS

Mercer, Joseph. "Teaching Graphing Concepts with Graphics Calculators." *Mathematics Teacher* (April 1995); pp. 268–273.

Smith, Cynthia Marie. "A Discourse on Discourse: Wrestling with Teaching Rational Equations." *Mathematics Teacher* (December 1998); pp. 749–753.

ACTIVITIES/MANIPULATIVES

Keller, Brian A. and Heather A. Thompson. "Whelk-come to Mathematics." *Activities: The Mathematics Teacher* (September 1999); pp. 475–489.

SOFTWARE

Harvey, Wayne and Judah L. Schwartz. *Visualizing Algebra: The Function Analyzer.* Pleasantville, NY; Sunburst Communications.

NAME ______________________ DATE __________

Parent Guide for Student Success

For use with Chapter 9

Chapter Overview One way that you can help your student succeed in Chapter 9 is by discussing the lesson goals in the chart below. When a lesson is completed, ask your student to interpret the lesson goals for you and to explain how the mathematics of the lesson relates to one of the key applications listed in the chart.

Lesson Title	*Lesson Goals*	*Key Applications*
9.1: Inverse and Joint Variation	Write and use inverse and joint variation models.	• Oceanography • Home Repair • Astronomy
9.2: Graphing Simple Rational Functions	Graph simple rational functions and use the graph of a rational function to solve real-life problems.	• Business • Lightning • Economics
9.3: Graphing General Rational Functions	Graph general rational functions and use the graphs to solve real-life problems.	• Manufacturing • Energy Expenditure • Hospital Costs
9.4: Multiplying and Dividing Rational Expressions	Multiply and divide rational expressions and use rational expressions to model real-life situations.	• Skydiving • Heat Generation • Farmland
9.5: Addition, Subtraction, and Complex Fractions	Add and subtract rational expressions. Simplify complex fractions.	• Statistics • Photography • Electronics
9.6: Solving Rational Equations	Solve rational equations and use rational equations to solve real-life problems.	• Chemistry • Football Statistics • Fuel Efficiency

Study Strategy

Making a Function Dictionary is the study strategy featured in Chapter 9 (see page 532). You may want to suggest that your student look back over the review material in previous chapters to identify the different types of functions to include in his or her function dictionary. Be sure your student includes the general form, a specific example, and a graph. If possible, provide graph paper to make it easier for your student to draw the graphs. Your student can add to this dictionary while working through Chapters 9–14.

NAME ______________________________ DATE ________

Parent Guide for Student Success

For use with Chapter 9

Key Ideas **Your student can demonstrate understanding of key concepts by working through the following exercises with you.**

Lesson	Exercise
9.1	The weight of an object varies inversely as the square of its distance from the center of Earth. Suppose that at sea level, about 4000 miles from the center of Earth, a person weighs 120 pounds. How much would the person weigh in the space shuttle 200 miles above sea level?
9.2	Describe the graph of the rational function $f(x) = \frac{3x - 1}{6x - 9}$.
9.3	Name the x-intercepts, vertical asymptotes, and horizontal asymptote(s) of the function $f(x) = \frac{x^2 + 4x - 5}{x^2 + 5x + 6}$.
9.4	A falling object accelerates until reaching terminal velocity. The greater the ratio of the object's volume to its cross-sectional surface area, the larger the terminal velocity. Suppose you need to drop an x-by-x-by-$3x$-meter box. Which option results in the lighter impact, dropping it with the x-by-$3x$ side down or with the square side down?
9.5	Perform the indicated operation and simplify. $\frac{x + 5}{x^2 + x - 6} - \frac{5}{x + 3}$
9.6	Elizabeth set a goal of saving an average of $10 per week. In the first six weeks, she was only able to save an average of $7 per week. How many weeks does Elizabeth need to save an average of $12 per week to raise the overall average to her goal of $10 per week?

Home Involvement Activity

You Will Need: Open trash can, ball of paper, tape measure
Directions: Measure 5 ft along the floor from the edge of a trash can. Stand at this distance and try to shoot the paper into the trash can. Make 10 tries. Record how many "baskets" you make. Repeat 3 times and find the average number of "baskets" you make. Find the average at 10 ft and 15 ft. The average number of baskets you make varies directly with the number you shoot and inversely with the distance you shoot from the basket. Write a variation model. Use the model to predict the average number of baskets you should make at 12 ft. Test your prediction.

Answers

9.1: about 109 lb **9.2:** *Sample answer:* The graph is a hyperbola through the points $(0, \frac{1}{9})$, $(2, \frac{5}{3})$, $(\frac{1}{3}, 0)$, and $(\frac{8}{3}, 1)$, with vertical asymptote $x = \frac{3}{2}$ and horizontal asymptote $y = \frac{1}{2}$.
9.3: x-intercepts: 1, −5; vertical asymptotes: $x = -3$ and $x = -2$; horizontal asymptote: $y = 1$
9.4: with the x-by-$3x$ side down **9.5:** $\frac{-4x + 15}{(x + 3)(x - 2)}$ **9.6:** 9 weeks

NAME ______________________________ DATE __________

Prerequisite Skills Review

For use before Chapter 9

EXAMPLE 1 Writing Direct Variation Equations

The variables x and y vary directly. Write an equation that relates the variables.

$x = 7, y = -3$

SOLUTION

Use the given values of x and y to find the constant of variation.

$y = kx$	Write direct variation equation.
$-3 = k(7)$	Substitute -3 for y and 7 for x.
$k = -\frac{3}{7}$	Solve for k.

The direct variation equation is $y = -\frac{3}{7}x$.

Exercises for Example 1

The variables x and y vary directly. Write an equation that relates the variables.

1. $x = -3, y = -6$
2. $x = 15, y = -5$
3. $x = -8, y = 1.6$

EXAMPLE 2 Multiplying Polynomials

Multiply the polynomials.

a. $-4(2x - 5)$

b. $x^2(x + 3)^2$

SOLUTION

a. $-4(2x - 5) = -8x + 20$ Distributive property

b. $x^2(x + 3)^2 = x^2((x)^2 + 2(3)(x) + (3)^2)$ Square of a binomial

$= x^2(x^2 + 6x + 9)$

$= x^4 + 6x^3 + 9x^2$ Distributive property

Exercises for Example 2

Multiply the polynomials.

4. $6(2x - 3)$
5. $-x(x^2 + 7)^2$
6. $x(x - 3)(x + 8)$

NAME ______________________________ DATE ____________

Prerequisite Skills Review

For use before Chapter 9

EXAMPLE 3 *Factoring Polynomials*

Factor the polynomial.

a. $x^2 - 11x + 28$ **b.** $5x^3 + 5$

SOLUTION

a. You want $x^2 - 11x + 28 = (x + m)(x + n)$ where $mn = 28$ and $m + n = -11$.

Factors of 28 (mn)	$-1, -28$	$1, 28$	$-2, -14$	$2, 14$	$-4, -7$	$4, 7$
Sum of factors (m + n)	-29	29	-16	16	-11	11

The table shows that $m = -4$ and $n = -7$. So, $x^2 - 11x + 28 = (x - 4)(x - 7)$.

b. $5x^3 + 5 = 5(x^3 + 1)$ Factor common monomial.

$= 5(x + 1)(x^2 - x + 1)$ Sum of two cubes.

Exercises for Example 3

Factor the polynomial.

7. $x^2 - 4x - 12$ **8.** $3x^2 + 7x - 20$ **9.** $6x^3 - 48$

EXAMPLE 4 *Finding Zeros of a Polynomial Function*

Find all the real zeros of the function.

$y = x^2 + 3x - 10$

SOLUTION

Use factoring to write the function in intercept form.

$y = x^2 + 3x - 10$

$= (x + 5)(x - 2)$

The zeros of the function are -5 and 2.

Exercises for Example 4

Find all real zeros of the function.

10. $y = x^2 - 7x - 30$ **11.** $y = x^2 + 8x + 12$

12. $y = x^3 - 2x^2 - 5x + 6$

CHAPTER 9

NAME ______________________________ DATE ____________

Strategies for Reading Mathematics

For use with Chapter 9

Using Operations with Numerical Fractions as Models

You can translate the techniques you have already learned for working with numerical fractions to working with rational expressions.

For example, think of the steps involved in adding $\frac{5}{6}$ and $\frac{3}{8}$.

1. Find a common denominator.
   $6 = 2 \cdot 3$ and $8 = 2^3$, so the least common denominator (LCD) is $2^3 \cdot 3 = 24$.

2. Rewrite the fractions with the common denominator.
   $\frac{5}{6} = \frac{5 \cdot 4}{6 \cdot 4} = \frac{20}{24}$ and $\frac{3}{8} = \frac{3 \cdot 3}{8 \cdot 3} = \frac{9}{24}$

3. Add the fractions and simplify, if possible.
   $\frac{20}{24} + \frac{9}{24} = \frac{29}{24}$ or $1\frac{5}{24}$

Now, apply the same steps to add $\frac{x}{2x - 6}$ and $\frac{5}{x^2 - 9}$.

1. Find a common denominator.
   $2x - 6 = 2(x - 3)$ and $x^2 - 9 = (x + 3)(x - 3)$ so the LCD is $2(x + 3)(x - 3)$.

2. Rewrite the fractions with the common denominator.
   $\frac{x}{2(x - 3)} \cdot \frac{x + 3}{x + 3} = \frac{x^2 + 3x}{2(x + 3)(x - 3)}$ and $\frac{5}{(x + 3)(x - 3)} \cdot \frac{2}{2} = \frac{10}{2(x + 3)(x - 3)}$

3. Add the fractions and simplify, if possible.
   $\frac{x^2 + 3x}{2(x + 3)(x - 3)} + \frac{10}{2(x + 3)(x - 3)} = \frac{x^2 + 3x + 10}{2(x + 3)(x - 3)}$

STUDY TIP
Using Numerical Fractions as Models
To understand how an operation is performed with rational expressions, write out an example of the same operation applied to numerical fractions. The technique for numerical fractions can be used as a model for the operation with rational expressions.

STUDY TIP
Simplifying Rational Expressions
To simplify a rational expression, first factor the numerator and denominator, and then cancel out the common factors between them. Be sure to cancel factors, not terms.

Questions

1. Find the LCD of each pair of fractions.

   a. $\frac{1}{60}, \frac{9}{28}$ **b.** $\frac{x}{24y}, \frac{9}{18xy}$ **c.** $\frac{3x - 2}{x^2 - 1}, \frac{x}{x^2 - 3x + 2}$

2. Add each pair of fractions in Question 1.

CHAPTER 9 CONTINUED

NAME ______________________ DATE ____________

Strategies for Reading Mathematics

For use with Chapter 9

Visual Glossary

The Study Guide on page 532 lists the key vocabulary for Chapter 9 as well as review vocabulary from previous chapters. Use the page references on page 532 or the Glossary in the textbook to review key terms from prior chapters. Use the visual glossary below to help you understand some of the key vocabulary in Chapter 9. You may want to copy these diagrams into your notebook and refer to them as you complete the chapter.

GLOSSARY

inverse variation (p. 534) Two variables x and y show inverse variation provided $y = \frac{k}{x}$ where k in a nonzero constant.

constant of variation (p. 534) The nonzero constant (usually denoted k) in an inverse variation equation $\left(y = \frac{k}{x}\right)$.

rational function (p. 540) A function of the form $f(x) = \frac{p(x)}{q(x)}$ where $p(x)$ and $q(x)$ are polynomials and $q(x) \neq 0$.

hyperbola (p. 540) The graph of a rational function, characterized by two symmetrical parts called branches.

simplified form of a rational expression (p. 554) A rational expression in which the numerator and denominator have no common factors (other than ± 1).

Graphing Rational Functions

The simplest type of rational function is inverse variation. For example, if x and y are the length and width of a rectangle with area 16, y varies inversely with x. Note that only the right-hand branch of the graph applies to the example because it is dealing with area.

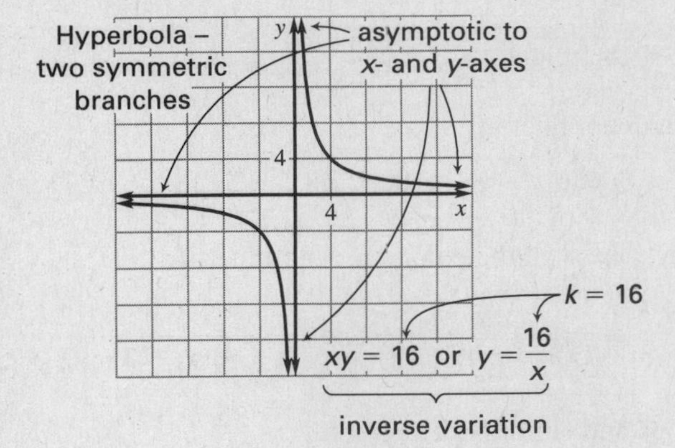

Simplifying Rational Expressions

To simplify a rational expression, factor the numerator and the denominator, and cancel out any common factors.

factor the numerator / factor the denominator; cancel out common factors

$$\frac{2x^3 - 32x}{x^3 + 8x^2 + 16x} = \frac{2\cancel{x}\cancel{(x+4)}(x-4)}{\cancel{x}\cancel{(x+4)}(x+4)}$$

$$= \frac{2(x-4)}{x+4}$$

x and 4 are terms, not factors. They cannot be cancelled.

LESSON 9.1

TEACHER'S NAME ____________ CLASS ______ ROOM ______ DATE ______

Lesson Plan

1-day lesson (See *Pacing the Chapter,* TE pages 530C–530D) **For use with pages 533–539**

GOALS
1. **Write and use inverse variation models.**
2. **Write and use joint variation models.**

State/Local Objectives ____________

✓ **Check the items you wish to use for this lesson.**

STARTING OPTIONS

____ Prerequisite Skills Review: CRB pages 5–6
____ Strategies for Reading Mathematics: CRB pages 7–8
____ Warm-Up or Daily Homework Quiz: TE pages 534 and 522, CRB page 11, or Transparencies

TEACHING OPTIONS

____ Concept Activity: SE page 533; CRB page 12 (Activity Support Master)
____ Lesson Opener (Application): CRB page 13 or Transparencies
____ Graphing Calculator Activity with Keystrokes: CRB pages 14–15
____ Examples 1–6: SE pages 534–536
____ Extra Examples: TE pages 535–536 or Transparencies
____ Closure Question: TE page 536
____ Guided Practice Exercises: SE page 537

APPLY/HOMEWORK

Homework Assignment

____ Basic 21–28, 30–42 even, 45–50, 59, 61–67 odd
____ Average 21–28, 30–44 even, 45–53, 59, 61–67 odd
____ Advanced 21–28, 30–44 even, 45–47, 54–60, 61–67 odd

Reteaching the Lesson

____ Practice Masters: CRB pages 16–18 (Level A, Level B, Level C)
____ Reteaching with Practice: CRB pages 19–20 or Practice Workbook with Examples
____ Personal Student Tutor

Extending the Lesson

____ Applications (Real-Life): CRB page 22
____ Challenge: SE page 539; CRB page 23 or Internet

ASSESSMENT OPTIONS

____ Checkpoint Exercises: TE pages 535–536 or Transparencies
____ Daily Homework Quiz (9.1): TE page 539, CRB page 26, or Transparencies
____ Standardized Test Practice: SE page 539; TE page 539; STP Workbook; Transparencies

Notes ____________

TEACHER'S NAME ______________________ CLASS ________ ROOM ________ DATE ________

Lesson Plan for Block Scheduling

Half-day lesson (See *Pacing the Chapter,* TE pages 530C–530D) **For use with pages 533–539**

1. Write and use inverse variation models.
2. Write and use joint variation models.

State/Local Objectives ______________________________

CHAPTER PACING GUIDE	
Day	**Lesson**
1	**9.1 (all)**; 9.2(all)
2	9.3 (all)
3	9.4 (all)
4	9.5 (all); 9.6(all)
5	Review/Assess Ch. 9

✓ Check the items you wish to use for this lesson.

STARTING OPTIONS

____ Prerequisite Skills Review: CRB pages 5–6
____ Strategies for Reading Mathematics: CRB pages 7–8
____ Warm-Up or Daily Homework Quiz: TE pages 534 and 522, CRB page 11, or Transparencies

TEACHING OPTIONS

____ Concept Activity: SE page 533; CRB page 12 (Activity Support Master)
____ Lesson Opener (Application): CRB page 13 or Transparencies
____ Graphing Calculator Activity with Keystrokes: CRB pages 14–15
____ Examples 1–6: SE pages 534–536
____ Extra Examples: TE pages 535–536 or Transparencies
____ Closure Question: TE page 536
____ Guided Practice Exercises: SE page 537

APPLY/HOMEWORK

Homework Assignment (See also the assignment for Lesson 9.2.)

____ Block Schedule: 21–28, 30–44 even, 45–53, 59, 61–67 odd

Reteaching the Lesson

____ Practice Masters: CRB pages 16–18 (Level A, Level B, Level C)
____ Reteaching with Practice: CRB pages 19–20 or Practice Workbook with Examples
____ Personal Student Tutor

Extending the Lesson

____ Applications (Real-Life): CRB page 22
____ Challenge: SE page 539; CRB page 23 or Internet

ASSESSMENT OPTIONS

____ Checkpoint Exercises: TE pages 535–536 or Transparencies
____ Daily Homework Quiz (9.1): TE page 539, CRB page 26, or Transparencies
____ Standardized Test Practice: SE page 539; TE page 539; STP Workbook; Transparencies

Notes ______________________________

LESSON 9.1

NAME ______________________ DATE __________

Available as a transparency

Warm-Up Exercises

For use before Lesson 9.1, pages 533–539

Solve for *y* in each equation.

1. $x + y = 2$
2. $xy = 8$
3. $2y = x^2$
4. $0.1 = xy$
5. $x = 8y$

Lesson 9.1

Daily Homework Quiz

For use after Lesson 8.8, pages 517–522

1. Evaluate $f(x) = \dfrac{5}{1 - 4e^{-2x}}$ for $f(2.5)$.
2. Graph $y = \dfrac{5}{1 + e^{-0.6x}}$. Identify the asymptotes, y-intercept, and point of maximum growth.
3. Solve the equation $\dfrac{15}{1 + 2e^{-3x}} = 2$.

NAME ______________________ DATE ____________

Activity Support Master

For use with page 533

Distance (m)	3	4	5	6	7	8	9
Apparent height (cm)							

LESSON **9.1**

NAME ______________________ DATE __________

Available as a transparency

Application Lesson Opener

For use with pages 534–539

Consider the relationship between the number of people helping to build a new house for a community and the number of hours it takes to build the house. (Assume that the people all work 8 hours per day and have comparable skills.)

1. If the number of people increases, what will happen to the amount of time?
2. If the number of people decreases, what will happen to the amount of time?
3. If the job deadline is pushed up, what can be done to get the job done in time?

Suppose your friends are purchasing a time-share vacation home. The cost for each person per week is $2000, divided by the number of people.

4. If the length of time is doubled, what will happen to the cost for each person?
5. If the number of people is tripled, what will happen to the cost for each person?
6. If the length of time and the number of people are *both* doubled, what will happen to the cost for each person?
7. Describe two different ways to reduce the cost for each person by half.

LESSON 9.1

NAME ______________________ DATE __________

Graphing Calculator Activity

For use with pages 534–539

GOAL **To use the graphing calculator to explore functions of inverse variation**

Two variables x and y show *inverse variation* if they are related as follows.

$y = \frac{k}{x}, k \neq 0$

Activity

1 Use a graphing calculator to graph $y = \frac{8}{x}$.

2 Use algebra to calculate the following.

a. Find y when $x = 2$.

b. Find y when $x = 3$.

c. Find x when $y = 4$.

d. Find x when $y = 8$.

3 Use the *Table* feature and the cursor keys to view different values of x and $y_1 = \frac{8}{x}$. Verify your answers in Step 2.

Exercises

1. Use a graphing calculator to graph $y = \frac{200}{x}$. Then use algebra to calculate the following.

a. Find y when $x = 5$.

b. Find y when $x = 6$.

c. Find x when $y = 100$.

d. Find x when $y = 200$.

2. Use the *Table* feature to verify your answers in Exercise 1.

3. Use a graphing calculator to graph $xy = 30$. Then use algebra to calculate the following.

a. Find y when $x = 5$.

b. Find y when $x = 8$.

c. Find x when $y = 3$.

d. Find x when $y = 5$.

4. Use the *Table* feature to verify your answers in Exercise 3.

NAME ______________________ DATE __________

Graphing Calculator Activity

For use with pages 534-539

TI-82

Step 1

Y= 8 ÷ X,T,θ ENTER ZOOM 6

Step 3

2nd [Tblset]1 ENTER 1 ENTER ENTER ▼ ENTER 2nd [TABLE]

TI-83

Step 1

Y= 8 ÷ X,T,θ,*n* ENTER ZOOM 6

Step 3

2nd [Tblset]1 ENTER 1 ENTER ENTER ▼ ENTER 2nd [TABLE]

SHARP EL-9600c

Step 1

Y= 8 ÷ X/θ/T/*n* ENTER ZOOM [A]5

Step 3

ENTER ▼ 1 ENTER 1 ENTER TABLE

CASIO CFX-9850GA PLUS

From the main manu, select GRAPH.

Step 1

8 ÷ X,θ,T EXE SHIFT F3 F3 EXIT F6

Step 3

MENU 7

F5 1 EXE 10 EXE 1 EXE

EXIT F6

LESSON 9.1

NAME ________________________ DATE ____________

Practice A

For use with pages 534–539

Tell whether x and y show *direct variation, inverse variation,* or *neither.*

1. $y = 3x$
2. $y = \frac{2}{x}$
3. $x + y = 7$
4. $xy = 5$

The variables x and y vary inversely. Use the given values to write an equation relating x and y. Then find y when x = 4.

5. $x = 2, y = 4$
6. $x = -3, y = 3$
7. $x = -4, y = 9$
8. $x = 3, y = 4$
9. $x = 16, y = \frac{1}{4}$
10. $x = 10, y = \frac{1}{2}$

Determine whether x and y show *direct variation, inverse variation,* or *neither.*

11.

x	y
3	12
8	32
11	44
0.5	2

12.

x	y
1	6
2	5
4	3
5	2

13.

x	y
3	1
6	0.5
10	0.3
12	0.25

14.

x	y
8	4
10	5
24	12
2	1

The variable z varies jointly with x and y. Use the given values to write an equation relating x, y, and z. Then find z when x = −3 and y = 4.

15. $x = 1, y = 2, z = 6$
16. $x = 2, y = 3, z = 4$
17. $x = 4, y = 3, z = 24$
18. $x = 8, y = -54, z = 144$

Simple Interest **In Exercises 19–21, use the following information.**

The simple interest I (in dollars) for a savings account is jointly proportional to the product of the time t (in years) and the principal P (in dollars). After six months, the interest on a principal of $2000 is $55.

19. Find the constant of variation k.
20. Write an equation that relates I, t, and P.
21. What will the interest be after two years?

NAME ______________________________ DATE __________

Practice B

For use with pages 534–539

Tell whether x and y show *direct variation, inverse variation,* or *neither.*

1. $x = \frac{y}{9}$

2. $y = \frac{1}{2}x$

3. $xy = 0.1$

4. $y = x + 5$

5.

x	y
5	15
8	24
1.5	4.5
0.5	1.5

6.

x	y
3	5
5	21
4.5	16.25
7	45

The variables x and y vary inversely. Use the given values to write an equation relating x and y. Then find y when x = 3.

7. $x = 6, y = 9$

8. $x = 72, y = \frac{1}{18}$

9. $x = 6, y = \frac{1}{2}$

The variable z varies jointly with x and y. Use the given values to write an equation relating x, y, and z. Then find z when x = 2 and y = −3.

10. $x = 2, y = 4, z = 6$

11. $x = 1, y = \frac{1}{8}, z = 4$

12. $x = \frac{1}{2}, y = 8, z = 12$

Simple Interest **In Exercises 13–15, use the following information.**

The simple interest I (in dollars) for a savings account is jointly proportional to the product of the time t (in years) and the principal P (in dollars). After nine months, the interest on a principal of \$3500 is \$91.88.

13. Find the constant of variation k.

14. Write an equation that relates I, t, and P.

15. What will the interest be after five years?

Boyle's Law **In Exercises 16–18, use the following information.**

Boyle's Law states that for a constant temperature, the pressure P of a gas varies inversely with its volume V. A sample of hydrogen gas has a volume of 8.56 cubic liters at a pressure of 1.5 atmospheres.

16. Find the constant of variation k.

17. Write an equation that relates P and V.

18. Find the volume of the hydrogen gas if the pressure changes to 1.2 atmospheres.

NAME ______________________ DATE ________

Practice C

For use with pages 534–539

Tell whether *x* and *y* show *direct variation, inverse variation,* or *neither.*

1. $x = 4y$

2. $x = \frac{5}{y}$

3. $x = \frac{y}{3}$

4. $\frac{2}{x} = \frac{7}{y}$

5.

x	y
1	4
2	2
0.5	8
0.25	16

6.

x	y
3	6
7	10
2.5	5.5
5.7	8.7

The variables *x* and *y* vary inversely. Use the given values to write an equation relating *x* and *y*. Then find *y* when $x = 6$.

7. $x = \frac{3}{2}, y = 12$

8. $x = 3, y = 0.1$

9. $x = \frac{1}{2}, y = \frac{2}{5}$

The variable *z* varies jointly with *x* and *y*. Use the given values to write an equation relating *x*, *y*, and *z*. Then find *z* when $x = 3$ and $y = -4$.

10. $x = 2, y = \frac{1}{8}, z = 3$

11. $x = \frac{3}{4}, y = \frac{5}{6}, z = 10$

12. $x = \frac{2}{3}, y = \frac{3}{4}, z = \frac{11}{12}$

Product Demand **In Exercises 13–15, use the following information.**

A company has found that the monthly demand d for one of its products varies inversely with the price p of the product. When the price is \$12.50, the demand is 12,000 units.

13. Find the constant of variation k.

14. Write an equation that relates d and p.

15. Find the demand if the price is reduced to \$12.00.

Specific Heat **In Exercises 16–18, use the following information.**

The amount of heat H (in kilocalories) necessary to change the temperature of an aluminum can is jointly proportional to the product mass m (in kilograms) and the temperature change desired T (in degrees Celsius). It takes 1.54 kilocalories of heat to change the temperature of a 0.028 kilogram aluminum can 250° C.

16. Find the constant of variation k.

17. Write an equation that relates H, m, and T.

18. How much heat is required to melt the can (at 660° C) if its current temperature is 20° C?

LESSON 9.1

NAME ______________________________ DATE __________

Reteaching with Practice

For use with pages 534–539

GOAL **Write and use inverse variation models and joint variation models**

VOCABULARY

Inverse variation is the relationship of two variables x and y if there is a nonzero number k such that $xy = k$, or $y = \frac{k}{x}$.

The nonzero constant k is called the **constant of variation.**

Joint variation occurs when a quantity varies directly as the product of two or more other quantities. For instance, if $z = kxy$ where $k \neq 0$, then z varies jointly with x and y.

EXAMPLE 1 *Classifying Direct and Inverse Variation*

Tell whether x and y show *direct variation*, *inverse variation*, or *neither*.

a. $x + y = 12$ **b.** $\frac{5}{y} = x$ **c.** $x = \frac{y}{2}$

SOLUTION

a. Because $x + y = 12$ cannot be rewritten in the form $y = kx$ or $y = \frac{k}{x}$, $x + y = 12$ shows neither type of variation.

b. If you multiply both sides of the equation $\frac{5}{y} = x$ by y, you obtain $xy = 5$.

When solving for y, the result is $y = \frac{5}{x}$, so $\frac{5}{y} = x$ shows inverse variation.

c. If you multiply both sides of the equation $x = \frac{y}{2}$ by 2, you obtain $y = 2x$, so $x = \frac{y}{2}$ shows direct variation.

Exercises for Example 1

Tell whether x and y show *direct variation, inverse variation,* or *neither.*

1. $xy = 8$ **2.** $y = x + 5$ **3.** $y = \frac{x}{2}$ **4.** $x = \frac{y}{3}$

EXAMPLE 2 *Writing an Inverse Variation Equation*

The variables x and y vary inversely, and $y = \frac{1}{2}$ when $x = 6$. Write an equation that relates x and y, and find y when $x = -3$.

NAME ______________________________ DATE ____________

Reteaching with Practice

For use with pages 534–539

Solution

Use the general equation for inverse variation to find k, the constant of variation.

$y = \frac{k}{x}$ Write general equation for inverse variation.

$\frac{1}{2} = \frac{k}{6}$ Substitute $\frac{1}{2}$ for y and 6 for x.

$3 = k$ Solve for k.

The inverse variation equation is $y = \frac{3}{x}$.

When $x = -3$, the value of y is:

$y = \frac{3}{-3} = -1$.

Exercises for Example 2

The variables *x* and *y* vary inversely. Use the given values to write an equation relating *x* and *y*. Then find *y* when *x* = 4.

5. $x = 10, y = 2$ **6.** $x = -3, y = 3$ **7.** $x = 2, y = 8$

EXAMPLE 3 *Writing a Joint Variation Model*

The variable z varies jointly with x and the square of y. When $x = 10$ and $y = 9$, $z = 135$. Write an equation relating x, y, and z, then find z when $x = 45$ and $y = 8$.

Solution

$z = kxy^2$ Write an equation for joint variation.

$135 = k(10)(9)^2$ Substitute 135 for z, 10 for x, and 9 for y.

$135 = 810k$ Simplify.

$\frac{1}{6} = k$ Solve for k.

The joint variation equation is $z = \frac{1}{6}xy^2$.

When $x = 45$ and $y = 8$:

$z = \frac{1}{6}(45)(8)^2 = 480$.

Exercises for Example 3

The variable *z* varies jointly with *x* and *y*. Use the given values to find an equation that relates the variables. Then find *z* when *x* = 2 and *y* = 8.

8. $x = 4, y = 3, z = 24$ **9.** $x = 8, y = -54, z = 144$ **10.** $x = 1, y = \frac{1}{8}, z = 4$

NAME ______________________________ DATE __________

Quick Catch-Up for Absent Students

For use with pages 533–539

The items checked below were covered in class on (date missed) ___________

Activity 9.1: Investigating Inverse Variation (p. 533)

____ **Goal:** Investigate inverse variation using the relationship between distance and height.

Lesson 9.1: Inverse and Joint Variation

____ **Goal 1:** Write and use inverse variation models. (pp. 534–535)

Material Covered:

____ Example 1: Classifying Direct and Inverse Variation

____ Example 2: Writing an Inverse Variation Equation

____ Example 3: Writing an Inverse Variation Model

____ Example 4: Checking Data for Inverse Variation

Vocabulary:

inverse variation, p. 534 constant of variation, p. 534

____ **Goal 2:** Write and use joint variation models. (p. 536)

Material Covered:

____ Student Help: Look Back

____ Example 5: Comparing Different Types of Variation

____ Example 6: Writing a Variation Model

Vocabulary:

joint variation, p. 536

____ Other (specify) ______________________________

____

Homework and Additional Learning Support

____ Textbook (specify) pp. 537–539

____

____ *Reteaching with Practice* worksheet (specify exercises)______________

____ *Personal Student Tutor* for Lesson 9.1

NAME ______________________ DATE ____________

Real-Life Application: When Will I Ever Use This?

For use with pages 534–539

Energy Efficient Housing Construction

The term *energy efficient house* basically means a house that uses energy as efficiently as possible. An energy efficient house consists of an airtight construction and high levels of insulation. With this simple idea, it is easy to build or remodel a house to become more energy efficient. These improvements typically add 5% to 10% to the total building or improvement costs. In return, a homeowner can expect to save 60% to 80% each year on energy costs.

Heat loss can occur through the ceiling, walls, windows, and doors. To find the amount of heat loss, you need to know the difference between the inside and the outside temperatures. During the cold season, the average interior temperature is 20° to 22° Celsius (68° to 72° Fahrenheit). The size of a heating system is based on the heat loss calculation using this interior temperature. You also need to know the surface area and the insulation levels. The table below shows the different RSI values for different insulation levels.

Insulation level	R-10	R-20	R-28	R-40	R-60
RSI value	1.8	3.5	4.9	7.0	10.6

In Exercises 1–5, use the following information.

The formula to calculate heat loss is:

$$\text{Heat Loss} = \frac{\text{Surface Area} \times \text{Difference in Temperature}}{\text{RSI Value}}$$

1. The interior temperature of a house is 20° Celsius and the outside temperature is 0° Celsius. The surface area of the ceiling is 84 square meters. Find the heat loss through the ceiling for the different levels of insulation. Organize your answers in a table.
2. Determine whether the heat loss from Exercise 1 is a *direct variation,* an *inverse variation,* or *neither.* Explain your answer.
3. A wall (without any doors or windows) has a surface area of 67 square meters. The insulation inside the walls is an R-28 level. The inside temperature is 20° Celsius. Complete the table to find the heat loss through this wall as the temperature drops from 10° Celsius to −10° Celsius.

Outside temperature	10° C	5° C	0° C	−5° C	−10° C
Heat loss					

4. A window is 3 feet high and 4 feet wide. The inside temperature is 20° Celsius and the outside temperature is −20° Celsius. The heat loss through this window is 960. Find the RSI value for this window.
5. Vertical blinds, with an RSI value of 0.1, are added to the window from Exercise 4. Will these vertical blinds increase or decrease the heat loss from this window? Explain your answer.

NAME ______________________ DATE __________

Challenge: Skills and Applications

For use with pages 534–539

1. The Ideal Gas Law states that, for an ideal gas, the pressure P, the volume V, and the temperature T are related by the equation

$$PV = \frac{m}{M}RT,$$

where m is the mass and M is the molecular weight of the gas (both constant if the same mass is considered), and R is a universal constant. Express the variation of each variable with respect to the other two variables.

a. P **b.** V **c.** T

In Exercises 2–4, describe the variational relationship between x and z and demonstrate this relationship algebraically.

2. x varies directly with y, and y varies inversely with z.

3. x varies inversely with y, and y varies inversely with z.

4. x varies jointly with y^2 and w, and y varies directly with z, while w varies inversely with z.

5. The weight of an object on a planet varies directly with the planet's mass and inversely with the square of the planet's radius. If all planets had the same density, the mass of the planet would vary directly with its volume, which equals $\frac{4}{3}\pi r^3$.

a. Use this information to find how the weight of an object w varies with the radius of the planet, assuming that all planets have the same density.

b. Earth has a radius of 6378 km, while Mercury (whose density is almost the same as Earth's) has a radius of 4878 km. If you weigh 125 lb on Earth, how much would you weigh on Mercury?

6. The graph of $y = \frac{1}{x}$ has an unusual property that can be illustrated in the diagram at the right. If

$$\frac{p}{q} = \frac{p'}{q'},$$

then the shaded regions have equal areas. Let $A_p{}^q$ represent the area under the graph between p and q.

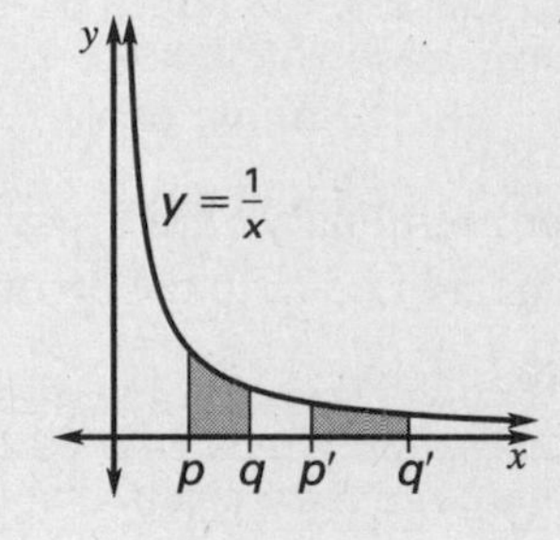

a. Use a diagram to show that $A_1{}^{ab} = A_1{}^a + A_a{}^{ab}$.

b. Use the the result of part (a) to show that $A_1{}^{ab} = A_1{}^a + A_1{}^b$.

LESSON 9.2

TEACHER'S NAME ______________________ CLASS ______ ROOM ______ DATE ______

Lesson Plan

1-day lesson (See *Pacing the Chapter,* TE pages 530C–530D) For use with pages 540–546

GOALS
1. Graph simple rational functions.
2. Use the graph of a rational function to solve real-life problems.

State/Local Objectives ____

____

✓ Check the items you wish to use for this lesson.

STARTING OPTIONS

____ Homework Check: TE page 537; Answer Transparencies
____ Warm-Up or Daily Homework Quiz: TE pages 540 and 539, CRB page 26, or Transparencies

TEACHING OPTIONS

____ Lesson Opener (Visual Approach): CRB page 27 or Transparencies
____ Graphing Calculator Activity with Keystrokes: CRB page 28
____ Examples 1–3: SE pages 541–542
____ Extra Examples: TE pages 541–542 or Transparencies; Internet
____ Technology Activity: SE page 546
____ Closure Question: TE page 542
____ Guided Practice Exercises: SE page 543

APPLY/HOMEWORK

Homework Assignment

____ Basic 12–28 even, 32–38 even, 41–43, 49–51, 53–67 odd
____ Average 12–28 even, 32–38 even, 41–46, 49–51, 53–69 odd
____ Advanced 12–40 even, 41–46, 49–52, 53–69 odd

Reteaching the Lesson

____ Practice Masters: CRB pages 29–31 (Level A, Level B, Level C)
____ Reteaching with Practice: CRB pages 32–33 or Practice Workbook with Examples
____ Personal Student Tutor

Extending the Lesson

____ Applications (Interdisciplinary): CRB page 35
____ Challenge: SE page 545; CRB page 36 or Internet

ASSESSMENT OPTIONS

____ Checkpoint Exercises: TE pages 541–542 or Transparencies
____ Daily Homework Quiz (9.2): TE page 545, CRB page 39, or Transparencies
____ Standardized Test Practice: SE page 545; TE page 545; STP Workbook; Transparencies

Notes ____

____

____

LESSON 9.2

TEACHER'S NAME ____________ CLASS ______ ROOM ______ DATE ______

Lesson Plan for Block Scheduling

Half-day lesson (See *Pacing the Chapter,* TE pages 530C–530D) **For use with pages 540–546**

GOALS
1. **Graph simple rational functions.**
2. **Use the graph of a rational function to solve real-life problems.**

State/Local Objectives ____________

CHAPTER PACING GUIDE	
Day	**Lesson**
1	9.1 (all); **9.2(all)**
2	9.3 (all)
3	9.4 (all)
4	9.5 (all); 9.6(all)
5	Review/Assess Ch. 9

✓ **Check the items you wish to use for this lesson.**

STARTING OPTIONS

____ Homework Check: TE page 537; Answer Transparencies
____ Warm-Up or Daily Homework Quiz: TE pages 540 and 539, CRB page 26, or Transparencies

TEACHING OPTIONS

____ Lesson Opener (Visual Approach): CRB page 27 or Transparencies
____ Graphing Calculator Activity with Keystrokes: CRB page 28
____ Examples 1–3: SE pages 541–542
____ Extra Examples: TE pages 541–542 or Transparencies; Internet
____ Technology Activity: SE page 546
____ Closure Question: TE page 542
____ Guided Practice Exercises: SE page 543

APPLY/HOMEWORK

Homework Assignment (See also the assignment for Lesson 9.1.)

____ Block Schedule: 12–28 even, 32–38 even, 41–46, 49–52, 53–69 odd

Reteaching the Lesson

____ Practice Masters: CRB pages 29–31 (Level A, Level B, Level C)
____ Reteaching with Practice: CRB pages 32–33 or Practice Workbook with Examples
____ Personal Student Tutor

Extending the Lesson

____ Applications (Interdisciplinary): CRB page 35
____ Challenge: SE page 545; CRB page 36 or Internet

ASSESSMENT OPTIONS

____ Checkpoint Exercises: TE pages 541–542 or Transparencies
____ Daily Homework Quiz (9.2): TE page 545, CRB page 39, or Transparencies
____ Standardized Test Practice: SE page 545; TE page 545; STP Workbook; Transparencies

Notes ____________

NAME ____________________ DATE ________

Available as a transparency

WARM-UP EXERCISES

For use before Lesson 9.2, pages 540–546

Tell how each equation is related to the graph of $y = \sqrt{x}$.

1. $y = \sqrt{x - 1}$

2. $y = \sqrt{x} + 1$

3. $y = -\sqrt{x}$

4. $y = \sqrt{x + 2} + 1$

5. $y = 2\sqrt{x}$

DAILY HOMEWORK QUIZ

For use after Lesson 9.1, pages 533–539

Tell whether x and y show *direct variation, inverse variation,* or *neither.*

1. $xy = \frac{2}{3}$

2. $y = 7x$

Write an equation for the given relation.

3. x varies inversely with y and $x = 2$ when $y = 12$.

4. z varies jointly with x and y and $z = 8$ when $x = 4$ and $y = 10$.

5. z varies inversely with the squares of x and directly with y.

NAME ______________________ DATE __________

Available as a transparency

Visual Approach Lesson Opener

For use with pages 540–545

When you studied quadratic functions, you learned that the graph of $y = (x - h)^2 + k$ has its vertex at (h, k).

For example, the graph of
$y = (x - 1)^2 + 2$
is the graph of $y = x^2$ shifted 1 unit to the right and 2 units up.

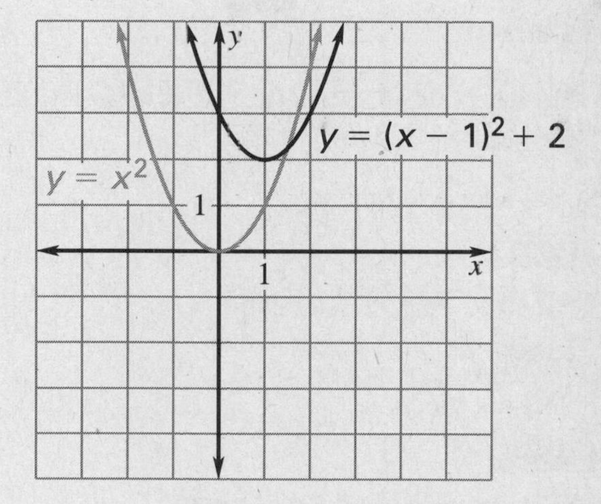

Likewise, the graph of $y = \dfrac{1}{x - 1} + 2$

is the graph of $y = \dfrac{1}{x}$ shifted 1 unit to the right and 2 units up.

In general, the graph of $y = \dfrac{1}{x - h} + k$

isthe graph of $y = \dfrac{1}{x}$ shifted h units to the right and k units up.

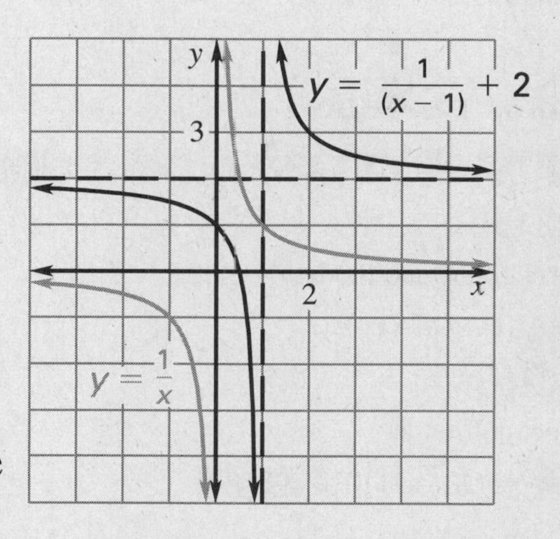

Graph each function.

1. $y = \dfrac{1}{x - 2} + 1$

2. $y = \dfrac{1}{x + 2} + 3$

3. $y = \dfrac{1}{x - 3} - 2$

4. $y = \dfrac{1}{x + 1} - 1$

LESSON 9.2

NAME ____________________ DATE ____________

Graphing Calculator Activity Keystrokes

For use with page 546

TI-82

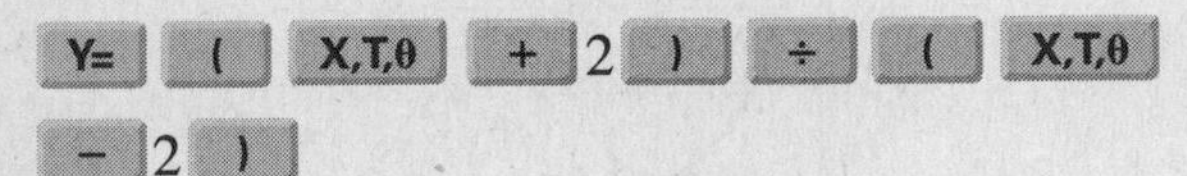

Select connected mode.

MODE

▼ ▼ ▼ ▼ ENTER

ZOOM 6

Select dot mode.

MODE

▼ ▼ ▼ ▼ ► ENTER

GRAPH

TI-83

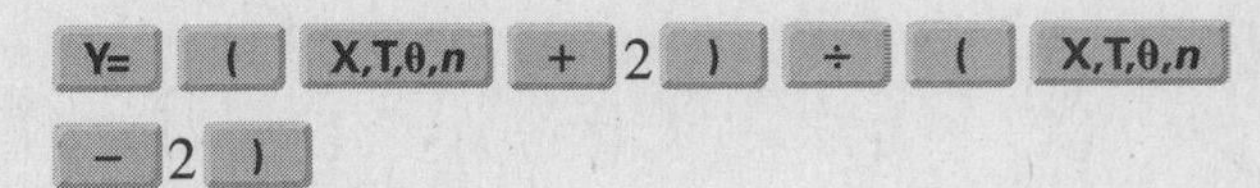

Select connected mode.

MODE

▼ ▼ ▼ ▼ ENTER

ZOOM 6

Select dot mode.

MODE

▼ ▼ ▼ ▼ ► ENTER

GRAPH

SHARP EL-9600C

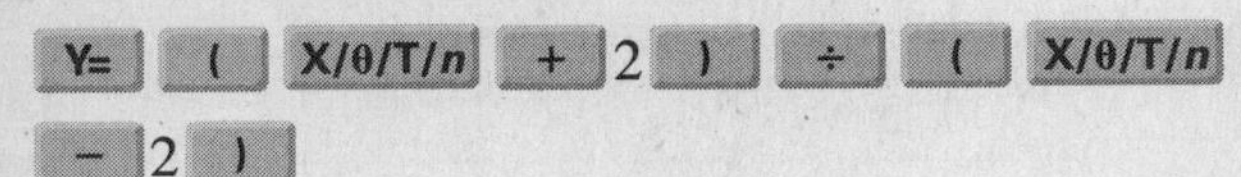

Select connected mode.

2ndF [FORMAT][E]1

ZOOM [A]5

Select dot mode.

2ndF [FORMAT][E]2 GRAPH

CASIO CFX-9850GA PLUS

From the main menu, choose GRAPH.

Select connected mode.

SHIFT [SET UP] F1 EXIT

SHIFT F3 F3 EXIT F6

Select dot mode.

SHIFT [SET UP] F2 EXIT F6

Lesson 9.2

LESSON 9.2

NAME ______________________________ DATE __________

Practice A

For use with pages 540–545

Find the domain of the function.

1. $f(x) = \dfrac{3}{x - 5}$
2. $f(x) = \dfrac{x + 4}{x + 6}$
3. $f(x) = \dfrac{2}{x} + 5$

Find the vertical asymptote of the graph of the function.

4. $f(x) = \dfrac{x}{x + 1}$
5. $f(x) = \dfrac{6}{x - 2} + 7$
6. $f(x) = \dfrac{x - 7}{x + 3}$

Find the horizontal asymptote of the graph of the function. Then state the range.

7. $f(x) = \dfrac{x - 3}{2x + 1}$
8. $f(x) = \dfrac{7}{x - 2} - 1$
9. $f(x) = \dfrac{4}{x} + 6$

Match the function with its graph.

10. $f(x) = \dfrac{2x + 1}{x - 1}$
11. $f(x) = \dfrac{x + 1}{x - 2}$
12. $f(x) = \dfrac{x - 1}{x + 2}$

A.

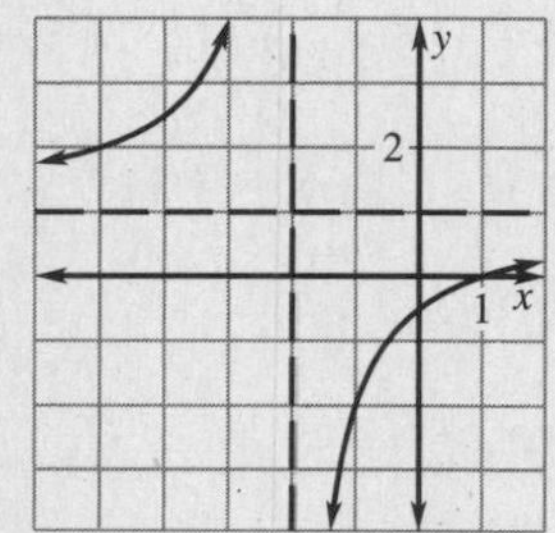

B.

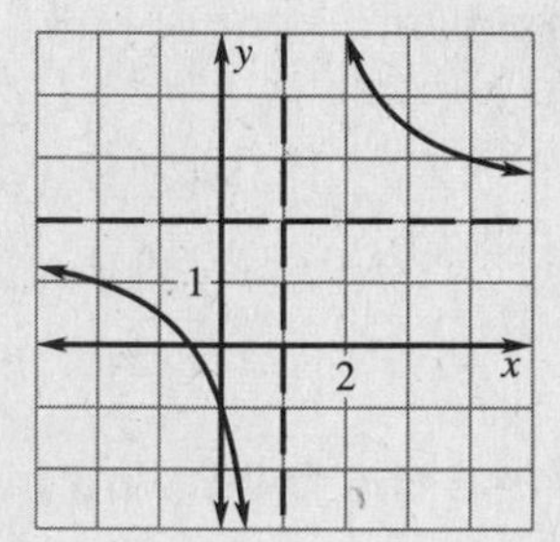

C.

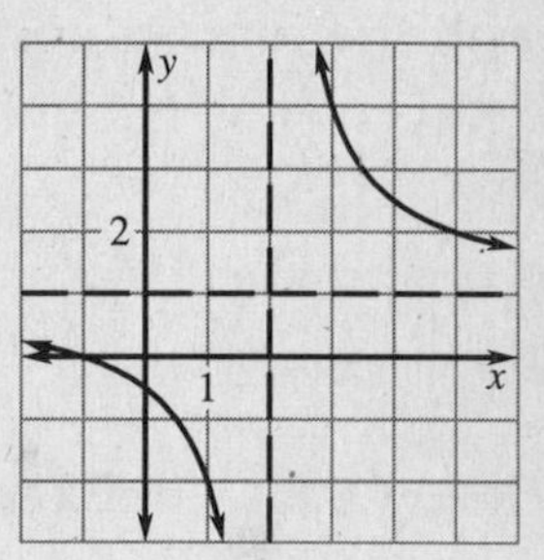

Graph the function.

13. $f(x) = \dfrac{x + 1}{x}$
14. $f(x) = \dfrac{3}{x - 2}$
15. $f(x) = \dfrac{3}{x + 2} + 2$

Sports Banquet **In Exercises 16–18, use the following information.**

You are organizing your high school's sports banquet. The banquet hall rental is \$250. In addition to this one-time charge, the meal will cost \$7 per plate. Let x represent the number of people who attend.

16. Write an equation that represents the total cost C.
17. Write an equation that represents the average cost A per person.
18. Graph the model in Exercise 17.

LESSON 9.2

NAME ______________________ DATE __________

Practice B

For use with pages 540–545

Identify the horizontal and vertical asymptotes of the graph of the function. Then state the domain and range.

1. $y = \dfrac{2}{x+4} - 5$

2. $y = \dfrac{3x-4}{4x+1}$

3. $y = \dfrac{2x+1}{3x-2} + 2$

Match the function with its graph.

4. $f(x) = \dfrac{2}{x-3} + 1$

5. $f(x) = \dfrac{2x-3}{x-3}$

6. $y = \dfrac{x+3}{x+2}$

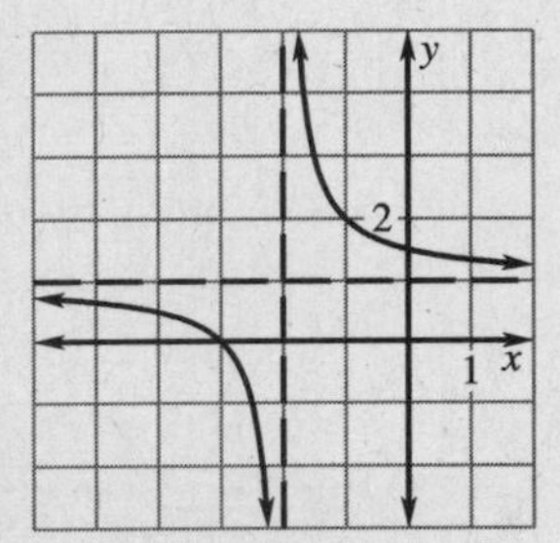

B.

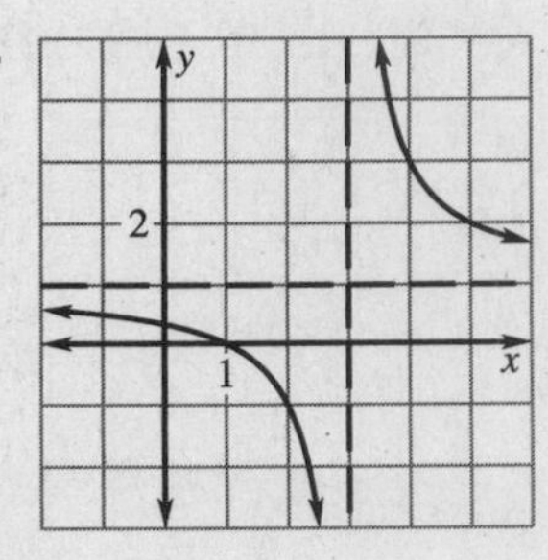

C.

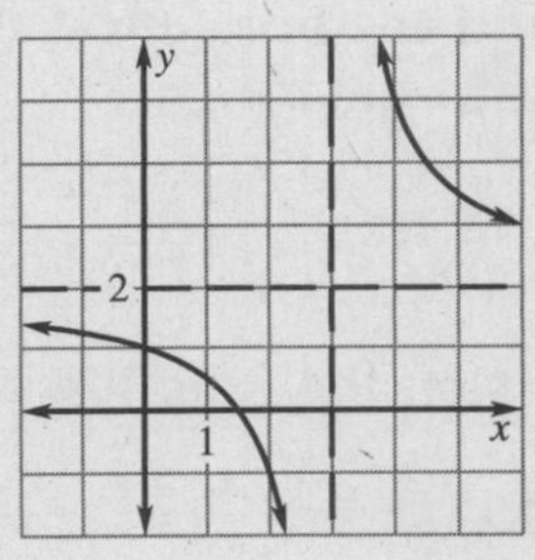

Graph the function. State the domain and range.

7. $y = -\dfrac{2}{x}$

8. $y = \dfrac{4}{x-2} + 3$

9. $y = -\dfrac{2}{x+3} - 1$

10. $y = \dfrac{x-1}{x-3}$

11. $y = \dfrac{3x-2}{-2x+3}$

12. $y = \dfrac{x}{2x-1}$

Inches of Rain **In Exercises 13–15, use the following information.**

The total number of inches of rain during a storm in a certain geographic area can be modeled by $r = \dfrac{2t}{t+8}$ where r is the amount of rain (in inches) and t is the length of the storm (in hours).

13. Graph the model.

14. What is an equation of the horizontal asymptote and what does the asymptote represent?

15. Use the graph to find the approximate number of inches of rain during a storm that lasts 5 hours.

LESSON
9.2

NAME ______________________________ DATE __________

Practice C

For use with pages 540–545

Identify the horizontal and vertical asymptotes of the graph of the function. Then state the domain and range.

1. $y = -\dfrac{1}{2x - 1} + 5$ **2.** $y = \dfrac{6x + 5}{-8x + 1}$ **3.** $y = \dfrac{12}{x + 7} - 10$

Match the function with its graph.

4. $f(x) = \dfrac{1}{x - 2} + 3$ **5.** $f(x) = -\dfrac{1}{x - 2} + 3$ **6.** $f(x) = \dfrac{x + 2}{x - 1}$

A.

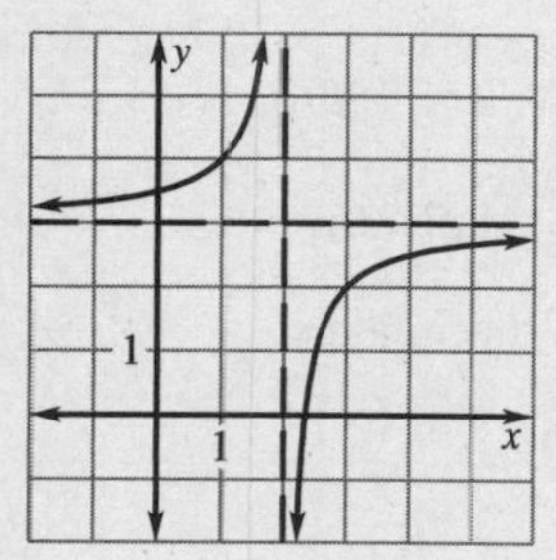

B.

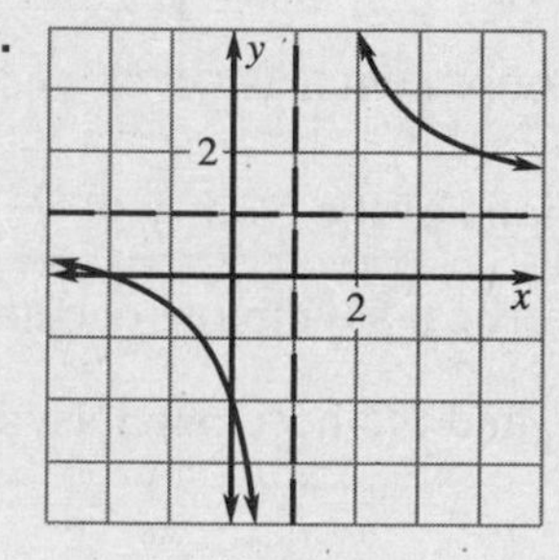

C.

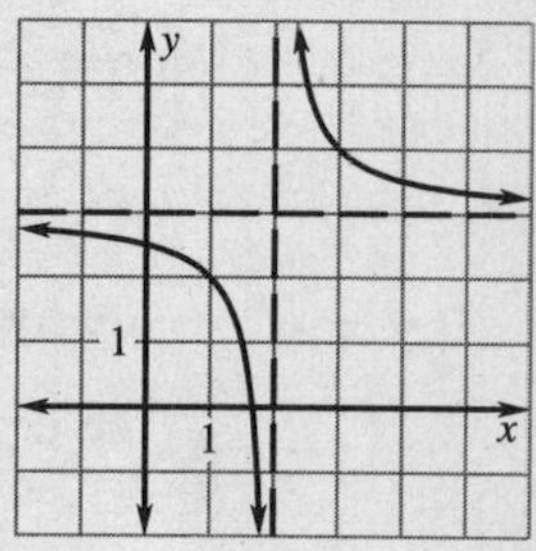

Graph the function. State the domain and range.

7. $y = \dfrac{4}{x} - 1$ **8.** $y = -\dfrac{2}{x - 3} + 4$ **9.** $y = \dfrac{2}{4x + 1} - 3$

10. $y = \dfrac{4x + 1}{2x - 3}$ **11.** $y = \dfrac{x - 5}{3x + 2}$ **12.** $y = \dfrac{5x}{-x - 3}$

Young's Rule **In Exercises 13–15, use the following information.**

Young's Rule is a formula that physicians use to determine the dosage levels of medicine for young children based on adult dosage levels. The child's dose can be modeled by $c = \dfrac{ta}{t + 12}$ where c is the child's dose (in milligrams), a is the adult's dose (in milligrams), and t is the age of the child (in years).

13. Graph the model for $t > 0$ and $a = 100$.

14. Use the graph to find the approximate dose for an eight-year-old child.

15. What is an equation of the horizontal asymptote and what does the asymptote represent?

LESSON 9.2

NAME ______________________ DATE __________

Reteaching with Practice

For use with pages 540–545

GOAL Graph simple rational functions and use the graph of a rational function to solve real-life problems

VOCABULARY

A **rational function** is a function of the form $f(x) = \frac{p(x)}{q(x)}$, where $p(x)$ and $q(x)$ are polynomials and $q(x) \neq 0$.

A **hyperbola** is the graph of a rational function of the form $f(x) = \frac{a}{x-h} + k$, whose center is (h, k), and asymptotes are $x = h$ and $y = k$. Rational functions of the form $y = \frac{ax+b}{cx+d}$ also have graphs that are hyperbolas. The vertical asymptote occurs at the x-value that makes the denominator zero, and the horizontal asymptote is the line $y = \frac{a}{c}$.

EXAMPLE 1 Graphing a Rational Function

Graph $y = \frac{3}{x-4} + 2$. State the domain and range.

SOLUTION

Begin by drawing the asymptotes $x = 4$ and $y = 2$. Then plot two points to the left of the vertical asymptote, such as $(3, -1)$ and $(1, 1)$, and two points to the right, such as $(5, 5)$ and $\left(6, \frac{7}{2}\right)$. Finally, use the asymptotes and plotted points to draw the branches of the hyperbola.

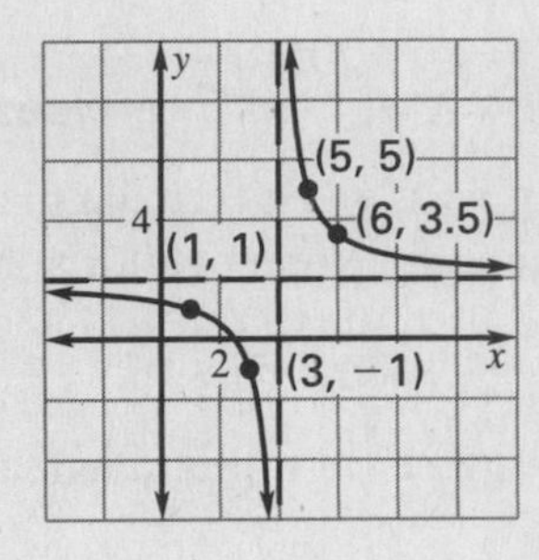

The domain is the set of real numbers except 4, because $x = 4$ will make the denominator zero. The range is the set of real numbers except $y = 2$, because that is where the horizontal asymptote occurs.

Exercises for Example 1

Graph the function. State the domain and range.

1. $y = \frac{-2}{x}$

2. $y = \frac{3}{x} + 5$

3. $y = \frac{2}{x-3} + 1$

4. $y = \frac{1}{x+5} - 2$

5. $y = \frac{-3}{x+2} - 1$

6. $y = \frac{2}{x-1} - 4$

EXAMPLE 2 Graphing a Rational Function

Graph $y = \frac{x-1}{x-3}$. State the domain and range.

NAME ______________________________ DATE __________

Reteaching with Practice

For use with pages 540–545

SOLUTION

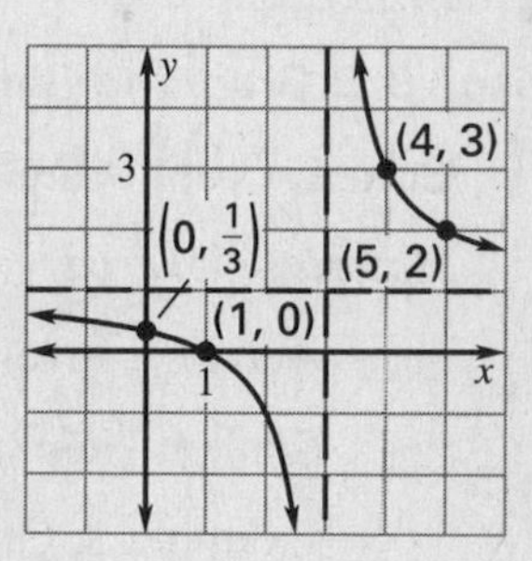

Begin by drawing the asymptotes. The vertical asymptote occurs at the x-value that makes the denominator zero, $x = 3$. The horizontal asymptote is the line $y = \frac{a}{c} = \frac{1}{1} = 1$. Then plot two points to the left of the vertical asymptote, such as $\left(0, \frac{1}{3}\right)$ and $(1, 0)$, and two points to the right, such as $(4, 3)$ and $(5, 2)$. Finally use the asymptotes and plotted points to draw the branches of the hyperbola. The domain is all real numbers except 3, because $x = 3$ makes the denominator zero. The range is all real numbers except 1, because $y = 1$ is a horizontal asymptote.

Exercises for Example 2

Graph the function. State the domain and range.

7. $y = \frac{x}{x + 2}$ **8.** $y = \frac{2x}{x - 4}$ **9.** $y = \frac{x + 2}{x + 3}$ **10.** $y = \frac{x - 1}{2x - 3}$

EXAMPLE 3 *Using a Rational Model*

The cost of cleaning up x percent of an oil spill that has washed ashore can be modeled by $C = \frac{20x}{101 - x}$, where C is the cost in thousands of dollars. Use a graph to approximate the cost to clean up 100% of the oil spill. Describe what happens to the cost as the percent cleanup increases.

SOLUTION

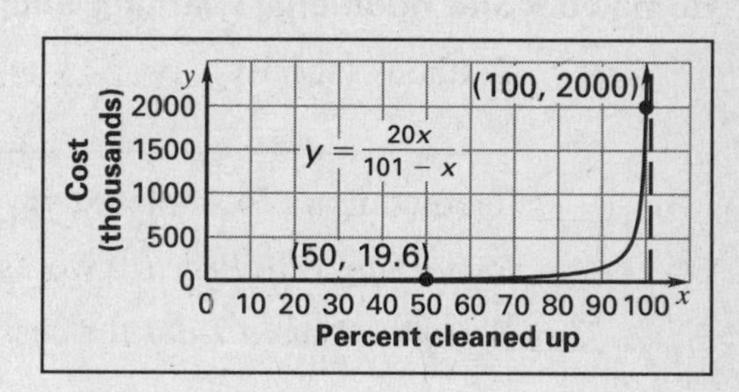

The graph of the model is shown at the right. A vertical asymptote occurs at $x = 101$. To clean up 100% of the oil spill it would cost approximately $2,000,000. Notice that the cleanup of 50% of the oil spill would cost only about $20,000. Therefore, the cost increases drastically as the percent cleanup approaches 100%.

Exercise for Example 3

11. The average cost C of producing x units can be modeled by

$$C = \frac{300{,}000 + 2x}{x}.$$

Use a graph to approximate the average cost of producing 10,000 units.

LESSON 9.2

NAME ______________________________ DATE __________

Quick Catch-Up for Absent Students

For use with pages 540–546

The items checked below were covered in class on (date missed) __________

Lesson 9.2: Graphing Simple Rational Functions

____ **Goal 1:** Graph simple rational functions. (pp. 540–541)

Material Covered:

____ Activity: Investigating Graphs of Rational Functions

____ Student Help: Look Back

____ Example 1: Graphing a Rational Function

____ Example 2: Graphing a Rational Function

Vocabulary:

rational function, p. 540

hyperbola, p. 540

branches, p. 540

____ **Goal 2:** Use the graph of a rational function to solve real-life problems. (p. 542)

Material Covered:

____ Example 3: Writing a Rational Model

Activity 9.2: Graphing Rational Functions (p. 546)

____ **Goal:** Use a graphing calculator to graph rational functions.

____ Student Help: Keystroke Help

____ Other (specify) ______________________________

____

Homework and Additional Learning Support

____ Textbook (specify) pp. 543–546

____

____ Internet: Extra Examples at www.mcdougallittell.com

____ *Reteaching with Practice* worksheet (specify exercises)__________

____ *Personal Student Tutor* for Lesson 9.2

Lesson 9.2

NAME ______________________________ DATE ____________

Interdisciplinary Application

For use with pages 540–545

Physics

GRAVITY In 1687, Isaac Newton developed his theory of gravity. This gravitational theory suggested a force between all objects. In addition he determined that this force does not require contact between the two objects and can act from a distance. Newton used his theories to determine that there is such a force between Earth and the moon, which keeps the moon in a circular motion around Earth. At the same time, there is an equal and opposite force which keeps the moon from orbiting any closer to Earth. These same theories can be applied to the motion of the planets (including Earth) around the sun. This classic theory, which is still accurate today, uses the Universal Constant of Gravitation (6.670×10^{-8}). The gravitational force between two objects can be determined using the formula:

$$\text{Gravitational Force} = (6.670 \times 10^{-8}) \times \frac{\text{Mass of 1st Object} \times \text{Mass of 2nd Object}}{(\text{Distance})^2}$$

1. The mass of Earth is 5.9736×10^{24} kilograms and the mass of the moon is 7.35×10^{22} kilograms. The moon is 384,400 kilometers from Earth. Find the gravitational force between Earth and the moon.

In Exercises 2–4, use the following information.

On December 18, 1999, NASA launched the first part of a series of satellites designed to monitor the climate and environmental changes on Earth. The mass of this satellite is 5190 kilograms. It will orbit Earth at an approximate altitude of 25,000 kilometers.

2. Write the function to find the gravitational force of this satellite as it is launched from Earth into the atmosphere.
3. Find the domain and the range of your function.
4. Graph your function from Exercise 2 for $0 < x \leq 25,000$. Describe what happens to the gravitational force as the satellite's distance from Earth increases.

LESSON 9.2

NAME ______________________ DATE ____________

Challenge: Skills and Applications

For use with pages 540–545

1. a. If $c = 0$ in the equation $y = \frac{ax + b}{cx + d}$, what kind of graph does the equation have?

   b. If $c \neq 0$ in the equation above, show that the equation can be written in the form $y = \frac{a'x + b'}{x + d'}$ for some constants a', b', and d'.

2. Find an equation of the hyperbola with asymptotes $x = 3$, $y = -1$, and passing through the point (5, 1).

3. Find an equation of the hyperbola with horizontal asymptote $y = 2$ and passing through the points $(-1, -4)$ and $\left(2, \frac{1}{2}\right)$.

4. Find an equation of the hyperbola passing through the points $(0, -3)$, $(1, 9)$, and $(2, 3)$.

5. By converting the form $y = \frac{ax + b}{cx + d}$ (with $c \neq 0$) to the form $y = \frac{p}{x - h} + k$, show that the asymptotes of the hyperbola are $x = -\frac{d}{c}$ and $y = \frac{a}{c}$. (*Hint*: Use the result of Exercise 1(b) above and long division.)

6. Show that as $x \to +\infty$, the corresponding y-values on the graph of $y = \frac{ax + b}{cx + d}$ (with $c \neq 0$) approach $\frac{a}{c}$. (*Hint*: Divide numerator and denominator by x.)

7. a. By plotting several points, determine the nature of the graph of $y = \frac{-3x + 6}{4x - 8}$.

   b. Write an equation involving a, b, c, and d that is true if and only if the graph of $y = \frac{ax + b}{cx + d}$ reduces to the kind of graph you found in part (a).

8. a. Assuming that the equation you wrote in Exercise 7(b) does *not* hold, write an equation for the inverse of the function $y = \frac{ax + b}{cx + d}$, in terms of a, b, c, and d.

   b. If the asymptotes of the original function are $x = h$, $y = k$, what are the asymptotes of the inverse function, in terms of h and k?

LESSON 9.3

TEACHER'S NAME ______________________ CLASS ________ ROOM ______ DATE ______

Lesson Plan

2-day lesson (See *Pacing the Chapter,* TE pages 530C–530D) For use with pages 547–553

GOALS
1. **Graph general rational functions.**
2. **Use the graph of a rational function to solve real-life problems.**

State/Local Objectives ____

____

✓ Check the items you wish to use for this lesson.

STARTING OPTIONS

____ Homework Check: TE page 543; Answer Transparencies
____ Warm-Up or Daily Homework Quiz: TE pages 547 and 545, CRB page 39, or Transparencies

TEACHING OPTIONS

____ Lesson Opener (Visual Approach): CRB page 40 or Transparencies
____ Examples: Day 1: 1–3: SE pages 547–548; Day 2: 4, SE page 549
____ Extra Examples: Day 1: TE pages 547–548 or Transp.; Day 2: TE page 549 or Transp.
____ Closure Question: TE page 549
____ Guided Practice: SE page 550 Day 1: Exs. 1–9; Day 2: Ex. 10

APPLY/HOMEWORK

Homework Assignment

____ Basic Day 1: 11–14, 20–25, 26–34 even; Day 2: 38–40, 47–48, 51–59 odd; Quiz 1: 1–13
____ Average Day 1: 11–16, 20–25, 26–34 even; Day 2: 38–41, 46–48, 51–59 odd; Quiz 1: 1–13
____ Advanced Day 1: 11–25, 26–36 even; Day 2: 38–41, 43–49, 51–59 odd; Quiz 1: 1–13

Reteaching the Lesson

____ Practice Masters: CRB pages 41–43 (Level A, Level B, Level C)
____ Reteaching with Practice: CRB pages 44–45 or Practice Workbook with Examples
____ Personal Student Tutor

Extending the Lesson

____ Applications (Real-Life): CRB page 47
____ Challenge: SE page 552; CRB page 48 or Internet

ASSESSMENT OPTIONS

____ Checkpoint Exercises: Day 1: TE pages 547–548 or Transp.; Day 2: TE page 549 or Transp.
____ Daily Homework Quiz (9.3): TE page 553, CRB page 52, or Transparencies
____ Standardized Test Practice: SE page 552; TE page 553; STP Workbook; Transparencies
____ Quiz (9.1–9.3): SE page 553; CRB page 49

Notes ____

____

____

LESSON 9.3

TEACHER'S NAME ____________ CLASS ______ ROOM ______ DATE ______

Lesson Plan for Block Scheduling

1-day lesson (See *Pacing the Chapter,* TE pages 530C–530D) **For use with pages 547–553**

GOALS
1. **Graph general rational functions.**
2. **Use the graph of a rational function to solve real-life problems.**

CHAPTER PACING GUIDE	
Day	**Lesson**
1	9.1 (all); 9.2(all)
2	**9.3 (all)**
3	9.4 (all)
4	9.5 (all); 9.6(all)
5	Review/Assess Ch. 9

State/Local Objectives ____________

____

____

✓ Check the items you wish to use for this lesson.

STARTING OPTIONS

____ Homework Check: TE page 543; Answer Transparencies

____ Warm-Up or Daily Homework Quiz: TE pages 547 and 545, CRB page 39, or Transparencies

TEACHING OPTIONS

____ Lesson Opener (Visual Approach): CRB page 40 or Transparencies

____ Examples: 1–4: SE pages 547–549

____ Extra Examples: TE pages 547–549 or Transparencies

____ Closure Question: TE page 549

____ Guided PracticeExercises: SE page 550

APPLY/HOMEWORK

Homework Assignment

____ Block Schedule: 11–16, 20–25, 26–34 even, 38–41, 46–48, 51–59 odd; Quiz 1: 1–13

Reteaching the Lesson

____ Practice Masters: CRB pages 41–43 (Level A, Level B, Level C)

____ Reteaching with Practice: CRB pages 44–45 or Practice Workbook with Examples

____ Personal Student Tutor

Extending the Lesson

____ Applications (Real-Life): CRB page 47

____ Challenge: SE page 552; CRB page 48 or Internet

ASSESSMENT OPTIONS

____ Checkpoint Exercises: TE pages 547– 549 or Transparencies

____ Daily Homework Quiz (9.3): TE page 553, CRB page 52, or Transparencies

____ Standardized Test Practice: SE page 552; TE page 553; STP Workbook; Transparencies

____ Quiz (9.1–9.3): SE page 553; CRB page 49

Notes ____________

____

____

Lesson 9.3

LESSON

9.3

NAME ______________________________ DATE __________

Available as a transparency

WARM-UP EXERCISES

For use before Lesson 9.3, pages 547–553

Find the solution(s) to each equation.

1. $(x - 3)(x + 3) = 0$
2. $(x - 4)(x + 1) = 0$
3. $x(x^2 - 1) = 0$
4. $x^2 - 4x - 5 = 0$
5. $x^2 + 1 = 0$

DAILY HOMEWORK QUIZ

For use after Lesson 9.2, pages 540–546

Identify the horizontal and vertical asymptotes of the graph of the function. State the domain and range. Graph the function.

1. $y = \dfrac{-3}{x + 1} + 2$
2. $y = \dfrac{x + 1}{2x + 3}$

Lesson 9.3

LESSON 9.3

NAME ______________________ DATE __________

Available as a transparency

Visual Approach Lesson Opener

For use with pages 547–553

You have learned how to graph simple rational functions. Many rational functions have more complicated graphs than the ones you have seen so far, but these graphs often have asymptotes similar to the ones you have seen.

For each graph below, sketch any vertical or horizontal asymptotes, and give their equations.

1. $y = \dfrac{x^2 + 1}{x^2 - 1}$

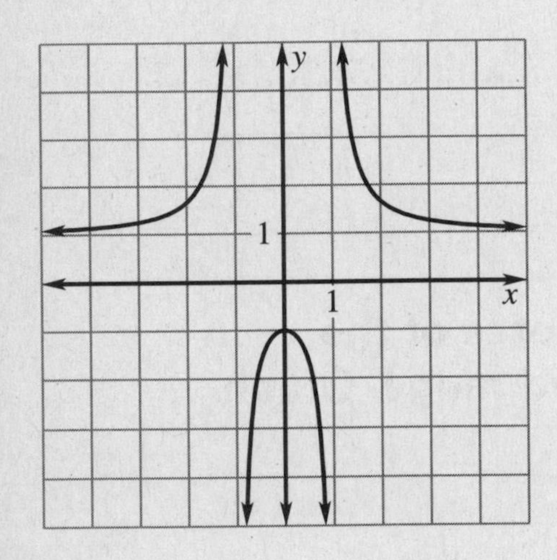

2. $y = \dfrac{2x^2}{x^2 + 1}$

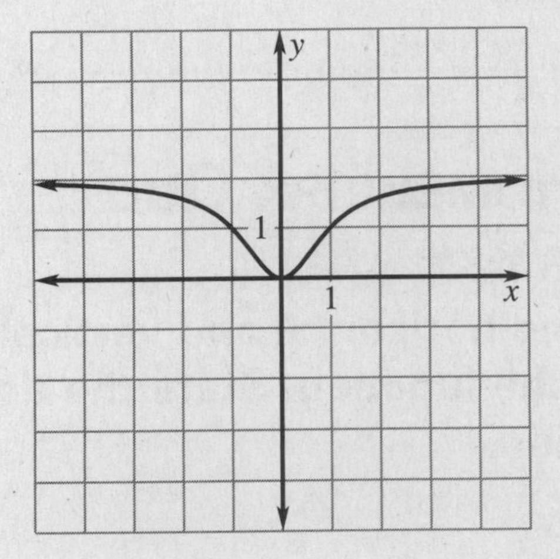

3. $y = \dfrac{4}{x^2 + x - 2}$

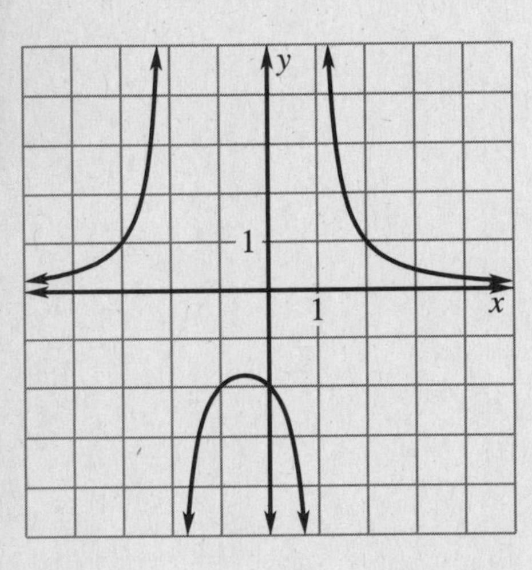

4. $y = \dfrac{x^2 + 4x + 6}{x + 2}$

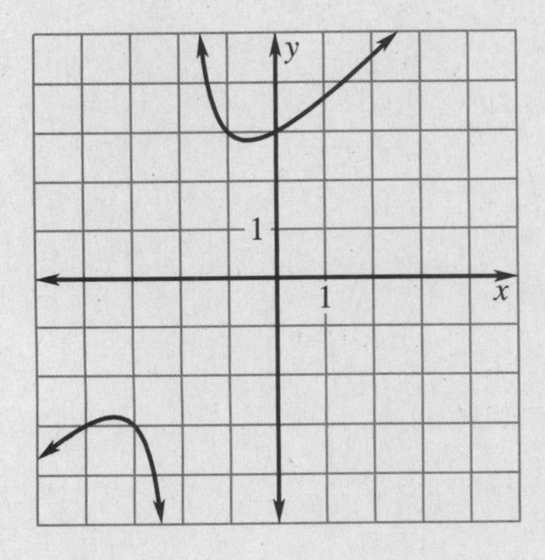

Lesson 9.3

LESSON **9.3**

NAME ______________________ DATE __________

Practice A

For use with pages 547–553

Identify the *x*-intercepts and vertical asymptotes of the graph of the function.

1. $y = \dfrac{x}{x^2 - 25}$

2. $y = \dfrac{x + 1}{x^2 - x - 6}$

3. $y = \dfrac{2x - 1}{x^2}$

4. $y = \dfrac{2}{x + 5}$

5. $y = \dfrac{x^2 - 8x - 9}{x^2 + 2}$

6. $y = \dfrac{x^2 - 6}{x}$

Match the function with its graph.

7. $y = \dfrac{x^2}{x^2 - 4}$

8. $y = \dfrac{5x}{x^2 + 4}$

9. $y = \dfrac{x^2 - 4x - 5}{x - 2}$

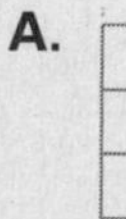

A.

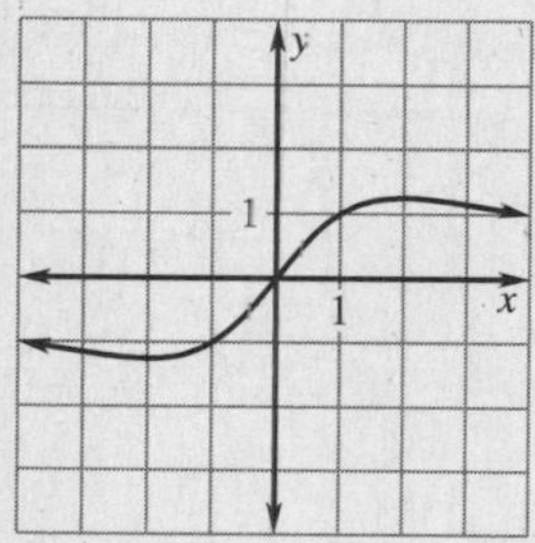

B.

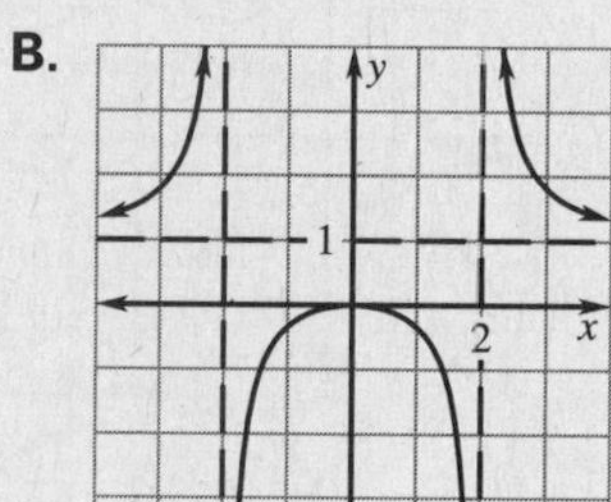

C.

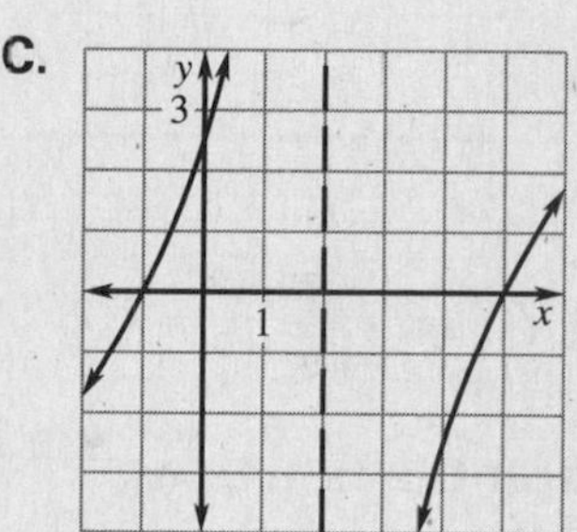

Graph the function.

10. $f(x) = \dfrac{x}{x^2 - 1}$

11. $f(x) = \dfrac{x^2 - 1}{x^2 - 4}$

12. $f(x) = \dfrac{x}{(x + 1)(x - 3)}$

13. $f(x) = \dfrac{x + 1}{x^2 - 4}$

14. $f(x) = \dfrac{x^2}{x + 1}$

15. $f(x) = \dfrac{x}{x^2 - 4x - 12}$

16. ***Garden Fencing*** Suppose you want to make a rectangular garden with an area of 450 square feet. You want to use the side of your house for one side of the garden and use fencing for the other three sides. Find the dimensions of the garden that minimize the length of fencing needed.

NAME ______________________________ DATE __________

Practice B

For use with pages 547–553

Identify the *x*-intercepts and vertical asymptotes of the graph of the function.

1. $y = \dfrac{2x^2 - 7x - 4}{x + 5}$

2. $y = \dfrac{x^2 + 1}{x^2 - 1}$

3. $y = \dfrac{3x^2 + 6x}{x^2 + 6x + 8}$

Match the function with its graph.

4. $y = -\dfrac{4}{x^2 - 5x + 4}$

5. $y = \dfrac{x^2 + 2}{x^2 - 16}$

6. $y = \dfrac{x^3}{x + 2}$

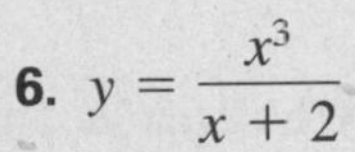

A.

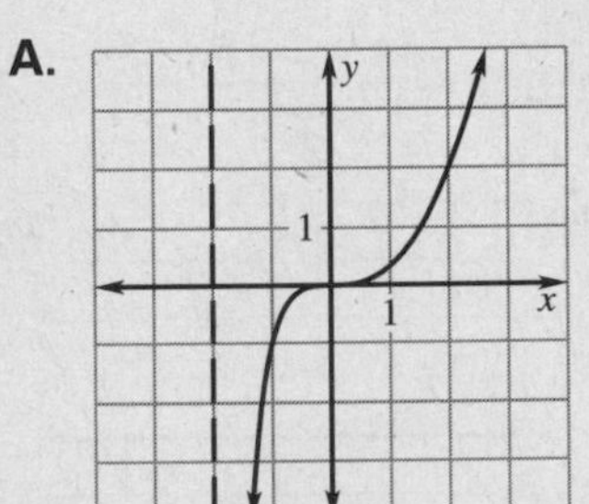

B.

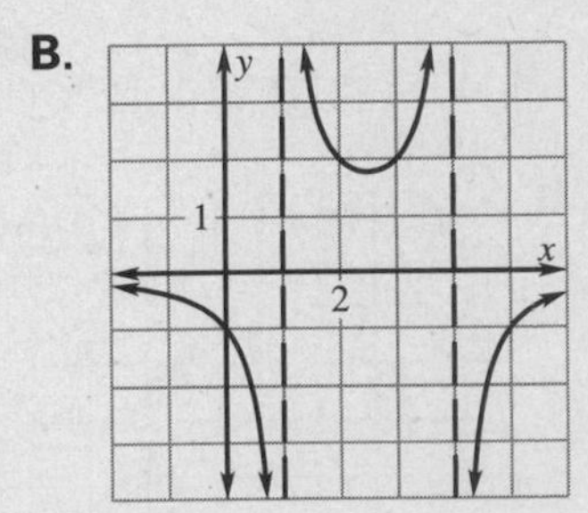

C.

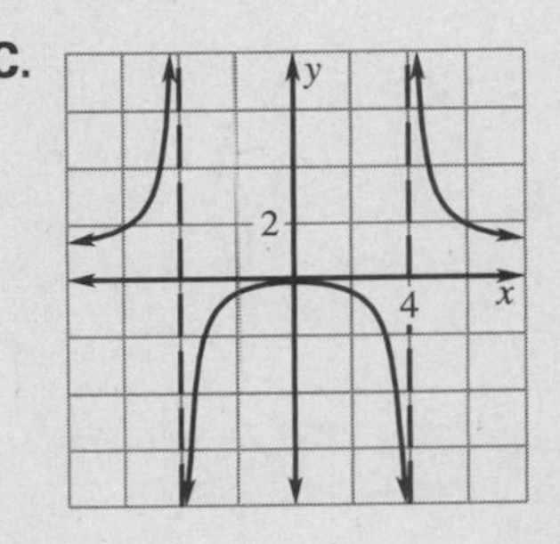

Graph the function.

7. $y = \dfrac{2x - 6}{x + 4}$

8. $y = \dfrac{5x + 1}{x^2 - 1}$

9. $y = \dfrac{3x^2 + 4x + 4}{x^2 - 5x - 6}$

10. $y = -\dfrac{3x^2}{2x + 6}$

11. $y = \dfrac{2x^2 + x - 9}{3x^2 - 12}$

12. $y = \dfrac{3x^2 - 1}{x^3}$

13. ***Critical Thinking*** Give an example of a rational function whose graph has two vertical asymptotes: $x = -3$ and $x = 0$, and one x-intercept: 5.

Pollution **In Exercises 14 and 15, use the following information.**

Suppose organic waste has fallen into a pond. Part of the decomposition process includes oxidation, whereby oxygen that is dissolved in the pond water is combined with decomposing material. Let $L = 1$ represent the normal oxygen level in the pond and let t represent the number of weeks after the waste is dumped. The oxygen level in the pond can be modeled by $L = \dfrac{t^2 - t + 1}{t^2 + 1}$.

14. Graph the model for $0 \le t \le 15$.

15. Explain how oxygen level changed during the 15 weeks after the waste was dumped.

LESSON 9.3

NAME ______________________________ DATE __________

Practice C

For use with pages 547–553

Identify the x-intercepts and vertical asymptotes of the graph of the function.

1. $y = \dfrac{5}{3x^2 - 7x - 6}$

2. $y = \dfrac{x^3 + 8}{x^2}$

3. $y = \dfrac{x^2 + 8x + 15}{x^2 + 4}$

Match the function with its graph.

4. $y = \dfrac{6}{x^2 - 9}$

5. $y = \dfrac{x^3 - 1}{x^2 + 2}$

6. $y = \dfrac{x^2 + 2x - 3}{2x^2 - x - 3}$

A.

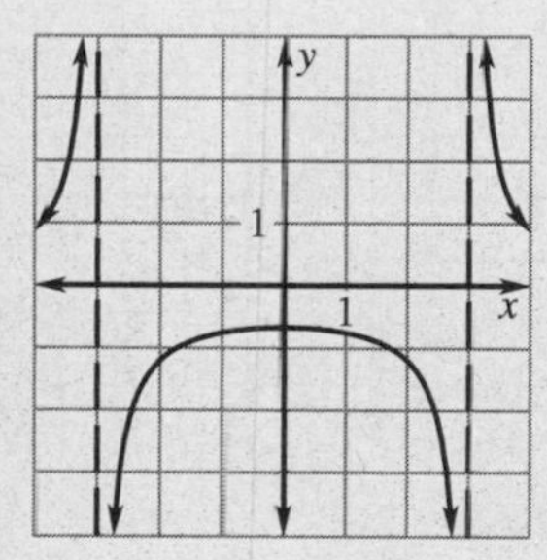

B.

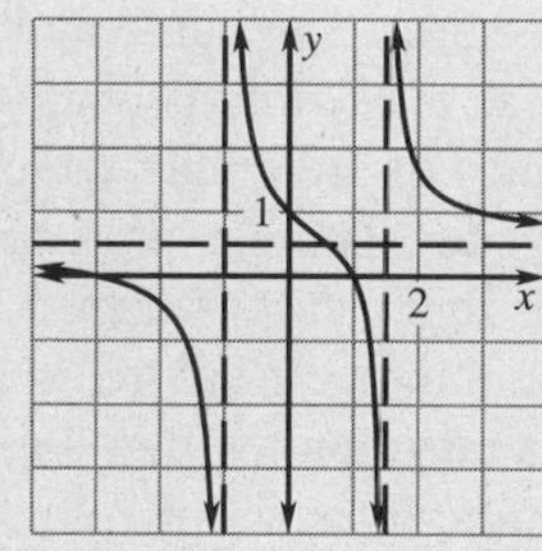

C.

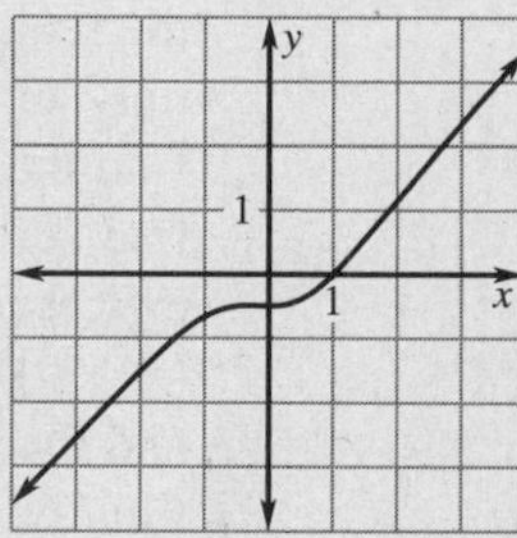

Graph the function.

7. $y = -\dfrac{2x^2}{x^2 - 9}$

8. $y = \dfrac{x^2 - 10x + 24}{3x}$

9. $y = \dfrac{x^2 - 6x + 9}{x^3 + 27}$

10. $y = \dfrac{x^2 - x - 2}{x - 1}$

11. $y = \dfrac{3x^3 + 1}{4x^3 - 32}$

12. $y = \dfrac{3}{4x + 10}$

13. ***Critical Thinking*** Give an example of a rational function whose graph has two vertical asymptotes: $x = 4$ and $x = -3$, one x-intercept: 0, and one horizontal asymptote: $y = 2$.

Manufacturing **In Exercises 14–17, use the following information.** A manufacturer of canned soup wants the volume of its cylindrical cans to be 300 cubic centimeters.

14. Use the volume formula $V = \pi r^2 h$ to express the can's height h as a function only of the can's radius r.

15. Use the surface area formula $S = 2\pi r^2 + 2\pi rh$ and your answer to Exercise 14 to express the can's surface area as a function only of the can's radius.

16. Graph the function from Exercise 15 on the domain $0 < r < 10$.

17. Find the dimensions of the can that has a volume of 300 cubic centimeters and uses the least amount of material possible.

LESSON 9.3

NAME ______________________ DATE ____________

Reteaching with Practice

For use with pages 547–553

GOAL **Graph general rational functions**

EXAMPLE 1 *Graphing a Rational Function* ($m < n$)

Graph $y = \dfrac{x}{x^2 - 4}$.

SOLUTION

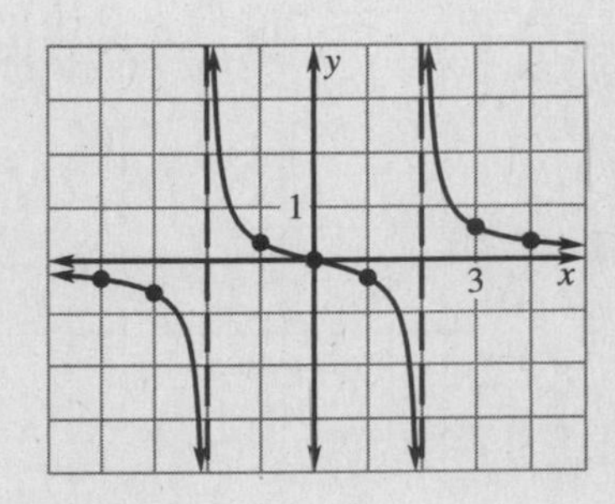

Begin by finding the x-intercepts of the graph, which are the real zeros of the numerator. The numerator has 0 as its only zero, so the graph has one x-intercept at $(0, 0)$. Then find the vertical asymptotes of the graph, which are the real zeros of the denominator. The denominator contains a difference of two squares that can be factored as $(x - 2)(x + 2)$, so the denominator has zeros 2 and -2, and the graph has vertical asymptotes $x = 2$ and $x = -2$. Finally, determine if the graph has a horizontal asymptote. Because the degree of the numerator (1) is less than the degree of the denominator (2), the line $y = 0$ is a horizontal asymptote. Construct a table of values consisting of x-values between and beyond the vertical asymptotes.

x	-4	-3	-1	0	1	3	4
y	$-\frac{1}{3}$	$-\frac{3}{5}$	$\frac{1}{3}$	0	$-\frac{1}{3}$	$\frac{3}{5}$	$\frac{1}{3}$

Plot the points and use the asymptotes to draw the graph.

Exercises for Example 1

Graph the function.

1. $y = \dfrac{2}{x + 1}$ **2.** $y = \dfrac{x}{x^2 - 1}$ **3.** $y = \dfrac{-3}{x - 2}$ **4.** $y = \dfrac{2x}{x^2 + 4}$

EXAMPLE 2 *Graphing a Rational Function* ($m = n$)

Graph $y = \dfrac{x^2 + 2}{x^2 - x - 6}$.

SOLUTION

The numerator has no zeros, so there are no x-intercepts. The denominator can be factored as $(x - 3)(x + 2)$, so the denominator has zeros 3 and -2 and the graph has vertical asymptotes $x = 3$ and $x = -2$. Because the degree of the numerator (2) equals the degree of the denominator (2), the line

$y = \dfrac{a_m}{b_n} = \dfrac{1}{1} = 1$ is a horizontal asymptote.

NAME ______________________ DATE ____________

Reteaching with Practice

For use with pages 547–553

Construct a table of values consisting of x-values between and beyond the vertical asymptotes.

x	-4	-3	-1	0	1	4	5
y	1.3	1.8	-0.75	-0.3	-0.5	3	1.9

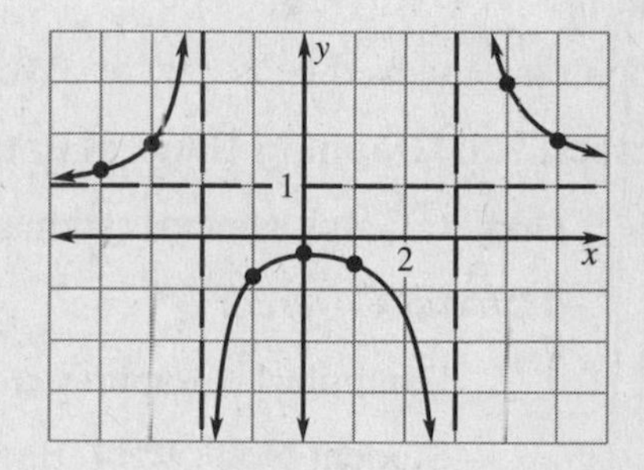

Plot the points and use the asymptotes to draw the graph.

Exercises for Example 2

Graph the function.

5. $y = \dfrac{x+1}{x-3}$ **6.** $y = \dfrac{2x+3}{x+2}$ **7.** $y = \dfrac{3x^2}{x^2+9}$ **8.** $y = \dfrac{2x^2}{x^2-1}$

EXAMPLE 3 Graphing a Rational Function ($m > n$)

Graph $y = \dfrac{x^3}{x^2-4}$.

SOLUTION

The numerator has 0 as its only zero, so the x-intercept of the graph is $(0, 0)$. The denominator can be factored as $(x+2)(x-2)$, so the graph has vertical asymptotes $x = 2$ and $x = -2$. The degree of the numerator (3) is greater than the degree of the denominator (2), so there is no horizontal asymptote. But, the end behavior of the graph is the same as the end behavior of the graph of $y = x^{3-2} = x$. So, the graph falls to the left and rises to the right. Construct a table of values consisting of x-values beyond the vertical asymptote.

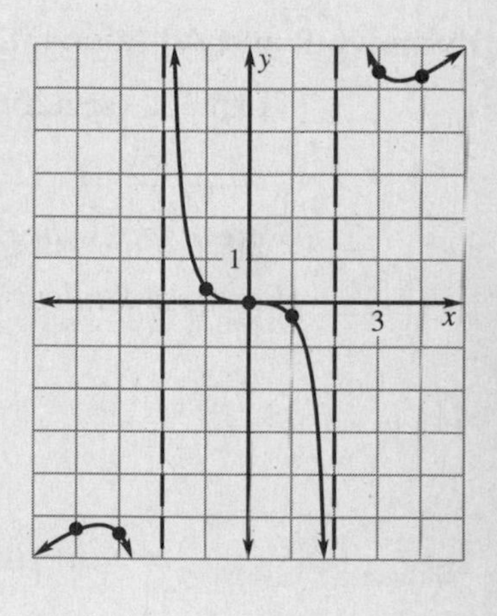

x	-4	-3	-1	1	3	4
y	-5.3	-5.4	0.3	-0.3	5.4	5.3

Plot the points, use the asymptotes, and consider the end behavior to draw the graph.

Exercises for Example 3

Graph the function.

9. $y = \dfrac{-x^2}{x+1}$

10. $y = \dfrac{x^2+1}{x-2}$

11. $y = \dfrac{x^2+3x-18}{x}$

12. $y = \dfrac{x^2-9}{2x}$

NAME ______________________________ DATE __________

Quick Catch-Up for Absent Students

For use with pages 547–553

The items checked below were covered in class on (date missed) __________

Lesson 9.3: Graphing General Rational Functions

____ **Goal 1:** Graph general rational functions. (pp. 547–548)

Material Covered:

____ Example 1: Graphing a Rational Function ($m < n$)

____ Student Help: Look Back

____ Example 2: Graphing a Rational Function ($m = n$)

____ Example 3: Graphing a Rational Function ($m > n$)

____ **Goal 2:** Use the graph of a rational function to solve real-life problems. (p. 549)

Material Covered:

____ Student Help: Look Back

____ Example 4: Finding a Local Minimum

____ Other (specify) ______________________________

____

Homework and Additional Learning Support

____ Textbook (specify) pp. 550–553

____

____ *Reteaching with Practice* worksheet (specify exercises)__________

____ *Personal Student Tutor* for Lesson 9.3

LESSON
9.3

NAME ______________________________ DATE __________

Real-Life Application: When Will I Ever Use This?

For use with pages 547–553

Marine Corps

In 1918 women were allowed to join the Marine Corps to handle the clerical duties, while the men fought in World War I. Prior to World War II, the United States Marine Corps Women's Reserve was formed. Eventually Congress passed the Women's Armed Services Integration Act. Women could then join the regular component of the Marine Corps, but the number of women could not exceed 2% of the total number of Marines.

By 1965, during the Vietnam War, there were 2700 women in the Marines. Women could participate in a wider variety of duties including radio operator, parachute rigger, quartermaster, auto mechanic, telegraph operator, and aerial gunnery instructor. But it was not until 1967 that a law was passed removing the restriction on the number of women in the Marines. In addition, women could be promoted to the rank of colonel. Within the next ten years, women would be able to work in all aspects of the Marine Corps except the infantry, artillery, armor, and the pilot/air crew.

In 1998 there were 9782 women in the Marine Corps on active duty, or about 5% of the Marines on active duty. Women now receive combat training and are accepted into special skill schools that were traditionally only available to the male Marines. Women are also working in nontraditional jobs and in some previously restricted areas. Women in the Marine Corps serve proudly and honorably next to their male counterparts in any area the United States Marine Corps requires.

The number of women in the United States Marine Corps can be modeled by

$$w = \frac{414.05x^2 - 3341.84x + 9403.15}{0.04x^2 - 0.32x + 1},$$ where x is the number of years since 1990.

1. Graph the model to find the number of women in the Marine Corps between 1990 and 1999.
2. Use your graph to estimate the number of female Marines in 1992, 1994, and 1996.
3. Identify the x-intercepts of this function, if any.
4. Identify the vertical and horizontal asymptotes of the graph of this function, if any.
5. Use the function to estimate the number of women in the Marine Corps in 2005.
6. Give two reasons why the number of women in the Marine Corps could exceed your answer from Exercise 5.

Lesson 9.3

NAME ______________________ DATE __________

Challenge: Skills and Applications

For use with pages 547–553

For Exercises 1–3, write a rational function satisfying the given conditions.

1. The graph of the function has vertical asymptotes $x = 3$ and $x = -3$ and horizontal asymptote $y = -2$.
2. The graph has vertical asymptotes at $x = 0$ and $x = 4$, and has horizontal asymptote $y = 0$.
3. The function has horizontal asymptote $y = 1$ and has a quadratic polynomial in its denominator, but has no vertical asymptote (i.e. its domain is all real numbers).
4. **a.** Graph the rational function $y = \dfrac{x^3 + 1}{x}$.

   b. On the same axes graph the functions $y = x^2$ and $y = \dfrac{1}{x}$. What do you notice about the relationship between these graphs and the graph of the function you drew in part (a)?
5. **a.** Graph the rational function $y = \dfrac{x^3}{x^2 - 9}$. Then graph the function $y = x$ on the same axes.

   b. What do you notice about the relationship between the graphs you drew in part (a)?
6. **a.** On separate axes, graph $y = \dfrac{1}{x - 2}$, $y = \dfrac{1}{(x - 2)^2}$, and $y = \dfrac{1}{(x - 2)^3}$.

   b. On the basis of your answer to part (a), make a conjecture about the y-values near the vertical asymptote of a function of the form $y = \dfrac{p(x)}{(x - a)^n}$, where n is an integer and $p(x)$ is a polynomial that does not have $x - a$ as a factor.
7. Find A and B in the following equation so that the two sides are equal for *all* values of x for which they are both defined:

$$\frac{A}{x - 2} + \frac{B}{x + 3} = \frac{3x - 11}{(x - 2)(x + 3)}.$$

(*Hint*: Recombine the two fractions on the left, and equate the coefficients of like terms in the numerators of both sides.)

LESSON 9.3

NAME ______________________ DATE __________

Quiz 1

For use after Lessons 9.1–9.3

Tell whether x and y show *direct variation, inverse variation,* or *neither*. *(Lesson 9.1)*

1. $\frac{y}{8} = x$

2. $xy = 16$

3. The variables x and y vary inversely. Write an equation relating x and y when $x = 5$ and $y = -3$. Then find y when $x = -5$. *(Lesson 9.1)*

4. The variable x varies jointly with y and z. Use the given values to write an equation relating x, y, and z when $x = 2$, $y = 3$, and $z = -4$. Then find y when $x = 8$ and $z = 1$. *(Lesson 9.1)*

Graph the function. *(Lessons 9.2 and 9.3)*

5. $y = \frac{-3}{x + 2} - 2$

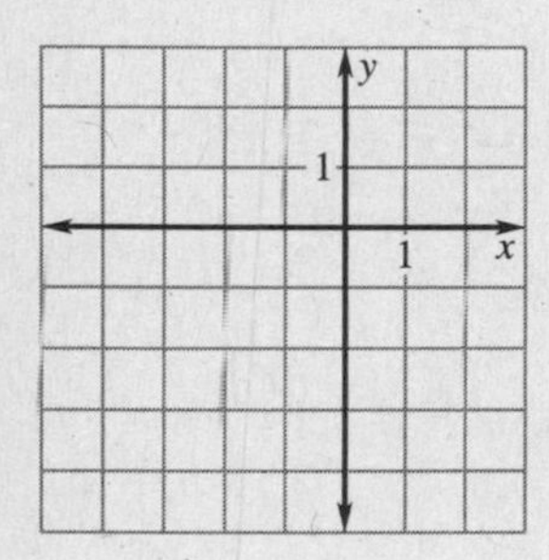

6. $y = \frac{x + 3}{3x - 1}$

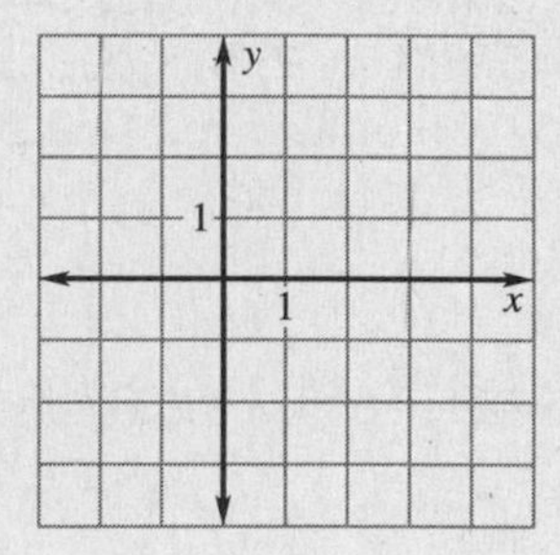

7. $y = \frac{5}{x^2 + 1}$

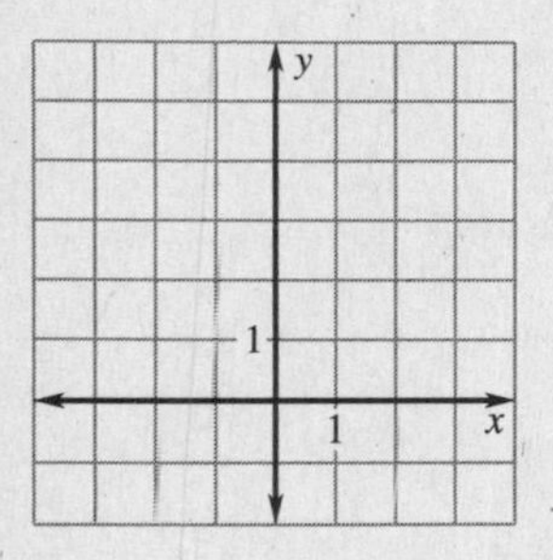

8. $y = \frac{x^2 + x - 6}{x + 2}$

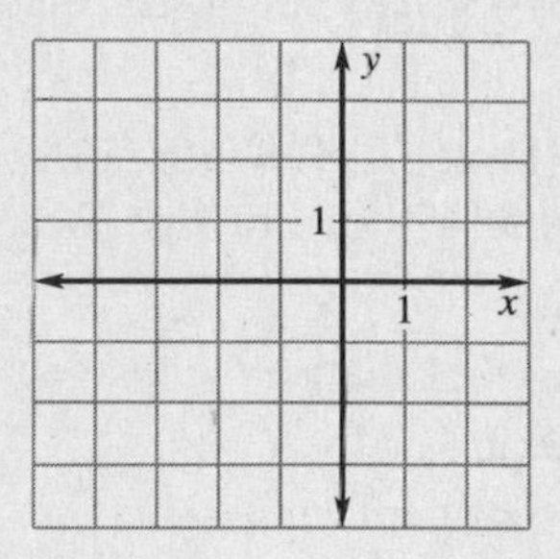

9. It is theorized that dividends paid on utilities stocks are inversely proportional to the prime interest rate. When the prime rate was 16%, dividends on a utility stock were $3.42 per share. If the prime rate R dropped to 11%, what dividends would be paid if the assumption of inverse proportionality is correct? *(Lesson 9.1)*

Answers

1. ______________
2. ______________
3. ______________
4. ______________
5. Use grid at left.
6. Use grid at left.
7. Use grid at left.
8. Use grid at left.
9. ______________

LESSON 9.4

TEACHER'S NAME ______________________ CLASS ________ ROOM ______ DATE ______

Lesson Plan

2-day lesson (See *Pacing the Chapter,* TE pages 530C–530D) **For use with pages 554–561**

1. Multiply and divide rational expressions.
2. Use rational expressions to model real-life quantities.

State/Local Objectives ____

____

✓ Check the items you wish to use for this lesson.

STARTING OPTIONS

____ Homework Check: TE page 550; Answer Transparencies
____ Warm-Up or Daily Homework Quiz: TE pages 554 and 553, CRB page 52, or Transparencies

TEACHING OPTIONS

____ Motivating the Lesson: TE page 555
____ Lesson Opener (Activity): CRB page 53 or Transparencies
____ Graphing Calculator Activity with Keystrokes: CRB page 54
____ Examples: Day 1: 1–5, SE pages 554–556; Day 2: 6–8, SE pages 556–557
____ Extra Examples: Day 1: TE pages 555–556 or Transp.; Day 2: TE pages 556–557 or Transp.; Internet
____ Technology Activity: SE page 561
____ Closure Question: TE page 557
____ Guided Practice: SE page 558 Day 1: Exs. 1–14; Day 2: Ex. 15

APPLY/HOMEWORK

Homework Assignment

____ Basic Day 1: 16–24, 28–33, 38–41; Day 2: 44–49, 54–57, 59, 63–73 odd
____ Average Day 1: 16–26, 28–35, 38–43; Day 2: 44–49, 54–59, 63–73 odd, 74–75
____ Advanced Day 1: 16–26, 28–35, 38–43; Day 2: 44–51, 54–59, 61–71 odd, 72–73

Reteaching the Lesson

____ Practice Masters: CRB pages 55–57 (Level A, Level B, Level C)
____ Reteaching with Practice: CRB pages 58–59 or Practice Workbook with Examples
____ Personal Student Tutor

Extending the Lesson

____ Applications (Interdisciplinary): CRB page 61
____ Challenge: SE page 560; CRB page 62 or Internet

ASSESSMENT OPTIONS

____ Checkpoint Exercises: Day 1: TE pages 555–556 or Transp.; Day 2: TE pages 556–557 or Transp.
____ Daily Homework Quiz (9.4): TE page 560, CRB page 65, or Transparencies
____ Standardized Test Practice: SE page 560; TE page 560; STP Workbook; Transparencies

Notes ____

____

____

LESSON
9.4

TEACHER'S NAME ______________________ CLASS ______ ROOM ______ DATE ______

Lesson Plan for Block Scheduling

1-day lesson (See *Pacing the Chapter,* TE pages 530C–530D) For use with pages 554–561

GOALS
1. **Multiply and divide rational expressions.**
2. **Use rational expressions to model real-life quantities.**

State/Local Objectives ______________________________

CHAPTER PACING GUIDE	
Day	**Lesson**
1	9.1 (all); 9.2(all)
2	9.3 (all)
3	**9.4 (all)**
4	9.5 (all); 9.6(all)
5	Review/Assess Ch. 9

✓ Check the items you wish to use for this lesson.

STARTING OPTIONS
____ Homework Check: TE page 550; Answer Transparencies
____ Warm-Up or Daily Homework Quiz: TE pages 554 and 553, CRB page 52, or Transparencies

TEACHING OPTIONS
____ Motivating the Lesson: TE page 555
____ Lesson Opener (Activity): CRB page 53 or Transparencies
____ Graphing Calculator Activity with Keystrokes: CRB page 54
____ Examples: 7–8: SE pages 554–557
____ Extra Examples: TE pages 555–557 or Transparencies; Internet
____ Technology Activity: SE page 561
____ Closure Question: TE page 557
____ Guided Practice Exercises: SE page 558

APPLY/HOMEWORK
Homework Assignment
____ Block Schedule: 16–26, 28–35, 38–43, 46–51, 54–59, 61–71 odd, 72–73

Reteaching the Lesson
____ Practice Masters: CRB pages 55–57 (Level A, Level B, Level C)
____ Reteaching with Practice: CRB pages 58–59 or Practice Workbook with Examples
____ Personal Student Tutor

Extending the Lesson
____ Applications (Interdisciplinary): CRB page 61
____ Challenge: SE page 560; CRB page 62 or Internet

ASSESSMENT OPTIONS
____ Checkpoint Exercises: TE pages 555–557 or Transparencies
____ Daily Homework Quiz (9.4): TE page 560, CRB page 65, or Transparencies
____ Standardized Test Practice: SE page 560; TE page 560; STP Workbook; Transparencies

Notes ______________________________

LESSON 9.4

NAME ______________________ DATE ________

Available as a transparency

WARM-UP EXERCISES

For use before Lesson 9.4, pages 554–561

Factor each expression.

1. $x^2 + 3x - 4$

2. $x^2 + 5x + 6$

3. $4x^2 - 9$

4. $6x^2 + x$

5. $8x^3 + 1$

DAILY HOMEWORK QUIZ

For use after Lesson 9.3, pages 547–553

Graph the function. Identify any asymptotes.

1. $y = \dfrac{x^2 + 2}{x - 1}$

2. $y = \dfrac{3x^2}{x^2 + 4}$

3. $y = \dfrac{2x + 7}{x^2 - 9}$

NAME ______________________ DATE __________

Available as a transparency

Activity Lesson Opener

For use with pages 554–560

SET UP: Work in groups of 3 or 4.

YOU WILL NEED: • index cards • pencil and paper

Write the following expressions on index cards (one per card) to create a set of 12 Expression Cards.

$\frac{10x - 5}{x^2 - 7x + 12}$	$\frac{x^2 - 5x + 4}{6x - 4}$	$\frac{3x^2 - x - 4}{6x - 2}$	$\frac{3x + 9}{x^2 + 3x - 4}$
$\frac{2x^2 + 9x + 9}{15x - 5}$	$\frac{3x^2 - 11x - 4}{x^2 + 3x}$	$\frac{x^2 + 6x + 8}{2x^2 - 5x + 2}$	$\frac{4x^2 + 3x - 1}{x^2 - 4x}$
$\frac{2x^2 - 7x - 4}{4x^2 + 12x + 9}$	$\frac{3x^2 + 8x + 4}{4x^2 - 4x - 3}$	$\frac{3x^2 + 11x + 6}{3x^2 + x - 4}$	$\frac{4x + 6}{4x^2 - 11x - 3}$

Write the following expressions on index cards (one per card) to create a set of 24 Factor Cards.

$x - 4$	$x - 3$	$x - 2$	$x - 1$	$x + 1$	$x + 2$
$x + 3$	$x + 4$	$2x - 3$	$2x - 1$	$2x + 1$	$2x + 3$
$3x - 4$	$3x - 2$	$3x - 1$	$3x + 1$	$3x + 2$	$3x + 4$
$4x - 1$	$4x + 1$	x	2	3	5

Place the Expression Cards face down in the center of the table. Shuffle the Factor Cards and deal all of them.

During each turn, an Expression Card is turned over and everyone works individually to factor the numerator and denominator of the expression. Any Factor Cards that match a factor in the expression are placed face up on the table. The winner is the first person to turn up all of his or her Factor Cards.

Factor carefully! If you turn up an incorrect card, you must take back all of your Factor Cards.

LESSON 9.4

NAME ______________________ DATE __________

Graphing Calculator Activity Keystrokes

For use with page 561

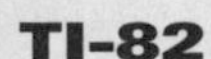

TI-82

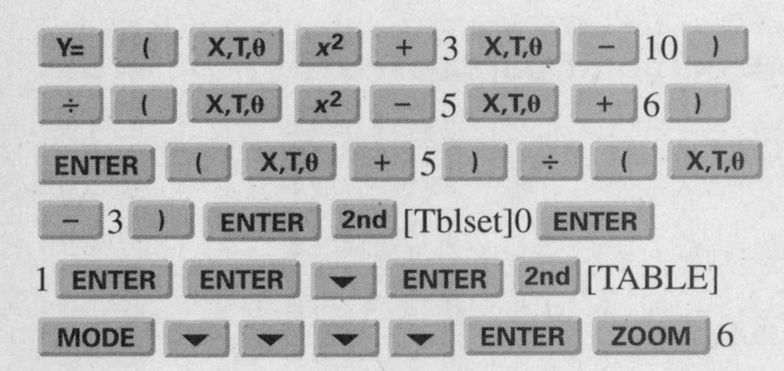

Y= (X,T,θ x2 + 3 X,T,θ − 10)
÷ (X,T,θ x2 − 5 X,T,θ + 6)
ENTER (X,T,θ + 5) ÷ (X,T,θ
− 3) ENTER 2nd [Tblset]0 ENTER
1 ENTER ENTER ▼ ENTER 2nd [TABLE]
MODE ▼ ▼ ▼ ▼ ENTER ZOOM 6

TI-83

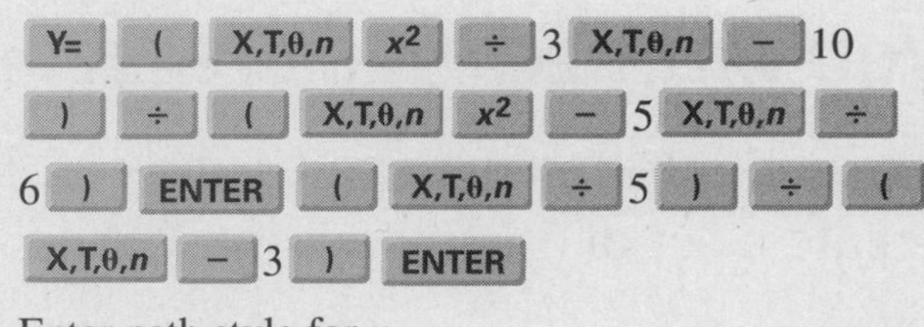

Y= (X,T,θ,n x2 ÷ 3 X,T,θ,n − 10
) ÷ (X,T,θ,n x2 − 5 X,T,θ,n ÷
6) ENTER (X,T,θ,n ÷ 5) ÷ (
X,T,θ,n − 3) ENTER

Enter path style for y_2.

Move the cursor to the left of y_2 and press ENTER until you get the path symbol.

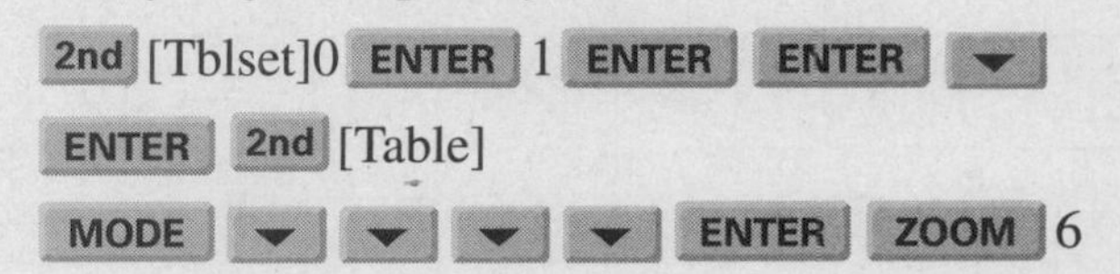

2nd [Tblset]0 ENTER 1 ENTER ENTER ▼
ENTER 2nd [Table]
MODE ▼ ▼ ▼ ▼ ENTER ZOOM 6

SHARP EL-9600c

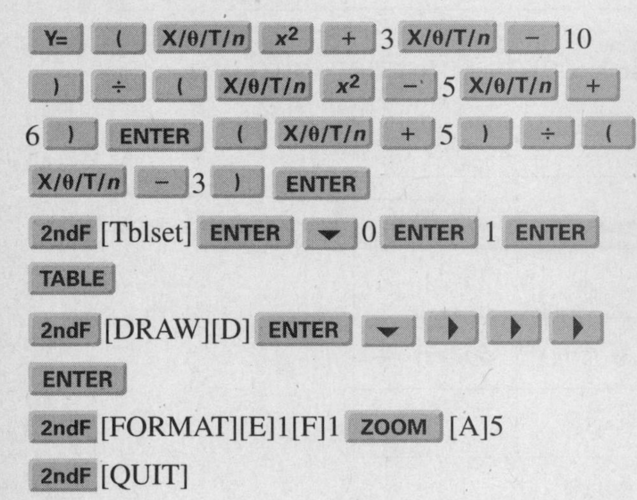

Y= (X/θ/T/n x2 + 3 X/θ/T/n − 10
) ÷ (X/θ/T/n x2 − 5 X/θ/T/n +
6) ENTER (X/θ/T/n + 5) ÷ (
X/θ/T/n − 3) ENTER
2ndF [Tblset] ENTER ▼ 0 ENTER 1 ENTER
TABLE
2ndF [DRAW][D] ENTER ▼ ▶ ▶ ▶
ENTER
2ndF [FORMAT][E]1[F]1 ZOOM [A]5
2ndF [QUIT]

CASIO CFX-9850GA PLUS

From the main menu, choose TABLE.

(X,θ,T x2 + 3 X,θ,T − 10) ÷
(X,θ,T x2 − 5 X,θ,T + 6) EXE
(X,θ,T + 5) ÷ (X,θ,T −
3) EXE F5 0 EXE 10 EXE 1 EXE EXIT F6
SHIFT F3 F3 EXIT F6 F5

Lesson 9.4

LESSON 9.4

NAME ____________________ DATE ________

Practice A

For use with pages 554–560

If possible, simplify the rational expression.

1. $\dfrac{4x^2}{2x^2 + 3x}$

2. $\dfrac{x^2 - 2x - 15}{x^2 - 4x - 5}$

3. $\dfrac{x^2 - 16}{x^2 + x - 12}$

4. $\dfrac{x^2 - 8x + 12}{x^2 + 3x - 10}$

5. $\dfrac{x^2 - 2x - 8}{x^2 + 3x - 4}$

6. $\dfrac{x^2 - 2x + 1}{x^2 - 1}$

Multiply the rational expressions. Simplify the result.

7. $\dfrac{12x^2y}{5y^2} \cdot \dfrac{2xy}{3x^2}$

8. $\dfrac{4y^2}{9x} \cdot \dfrac{27}{16xy^2}$

9. $\dfrac{x^2 - 2x}{x^2 + 2x + 1} \cdot \dfrac{x^2 + 4x + 3}{x^2 + 3x}$

10. $\dfrac{x^2 + 2x - 3}{x + 2} \cdot \dfrac{x^2 + 2x}{x^2 - 1}$

Divide the rational expressions. Simplify the result.

11. $\dfrac{5x^5}{8} \div \dfrac{15x^2}{12}$

12. $\dfrac{48x^2}{y} \div \dfrac{36xy^2}{5}$

13. $\dfrac{x^2}{x^2 - 1} \div \dfrac{3x}{x + 1}$

14. $\dfrac{x^2 - 9x - 22}{x^2 + 5x - 24} \div \dfrac{x + 2}{x - 3}$

Perform the indicated operations. Simplify the result.

15. $\dfrac{5x^2y}{2xy} \cdot \dfrac{6x^3y^5}{10y} \cdot \dfrac{3x}{y^3}$

16. $\dfrac{x - 11}{2x + 10} \div \dfrac{x^2 - 8x - 33}{x + 5} \cdot \dfrac{x + 3}{x^2}$

17. $(x^2 + x - 30) \div \dfrac{x^2 - 11x + 30}{x^2 + 7x + 12} \cdot \dfrac{x - 6}{x + 6}$

18. $\dfrac{x^2 - 5x - 14}{x^2 - 6x - 7} \cdot (x^2 - 4x - 5) \div \dfrac{x^2 + x - 30}{2}$

19. ***CDs and Cassettes*** Use the diagrams below to find the ratio of the volume of the compact disc storage crate to the volume of the cassette storage crate.

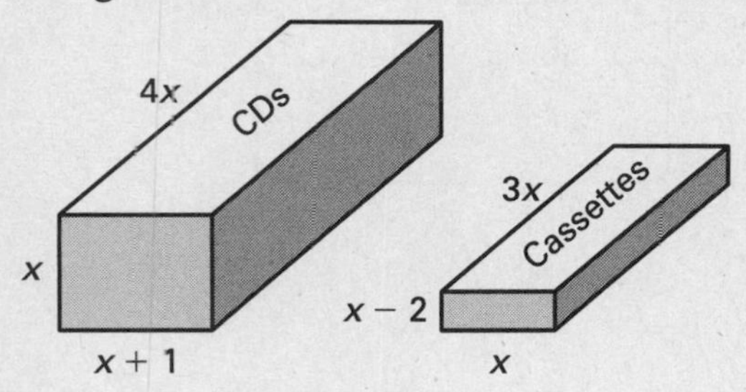

Lesson 9.4

NAME ____ DATE ____________

Practice B

For use with pages 554–560

If possible, simplify the rational expression.

1. $\dfrac{x^2 - 8x - 9}{x^2 - 1}$

2. $\dfrac{x + 3}{x^2 + 5x + 6}$

3. $\dfrac{x^2 - 4}{x^2 + 4}$

Multiply the rational expressions. Simplify the result.

4. $\dfrac{4x^2y^3}{x^5y^6} \cdot \dfrac{xy}{20x^3}$

5. $\dfrac{81x^9}{y^4} \cdot \dfrac{x^2}{36x^5y}$

6. $\dfrac{x^2 + 4x - 12}{x^4 + 9x^3 + 18x^2} \cdot 6x^2$

7. $\dfrac{3x^2 - 12}{5x - 10} \cdot \dfrac{1}{2x + 4}$

Divide the rational expressions. Simplify the result.

8. $\dfrac{12x^2y}{5y^2} \div \dfrac{3x^2}{2xy}$

9. $\dfrac{x^2 - 3x + 2}{25x} \div \dfrac{x - 1}{5x^2}$

10. $\dfrac{5x^2 - 20}{25x^2} \div \dfrac{x^2 + 6x + 8}{x^2 + 10x + 24}$

11. $(x + 7) \div \dfrac{x^2 + 9x + 14}{x^2 + 5x + 6}$

Perform the indicated operations. Simplify the result.

12. $(x^2 + x - 30) \div \dfrac{x^2 - 2x - 15}{x^2 + 7x + 12} \cdot \dfrac{x - 5}{x + 6}$

13. $\dfrac{x^2 + x - 20}{x + 1} \div \dfrac{33x^2 - 132x}{16x + 16} \div \dfrac{8x + 40}{11x + 44}$

14. $\dfrac{x^2 + 6x - 7}{3x^2} \cdot \dfrac{6x}{x + 7} \div \dfrac{x - 1}{4}$

15. $\dfrac{3xy^3}{x^3y} \cdot \dfrac{y}{6x} \div \dfrac{9y^2}{xy}$

Geometry Find the ratio of the area of the shaded region to the total area. Write your result in simplified form.

16.

7

$x + 1$

$x^2 + 5x$

$6(x + 1)$

17.

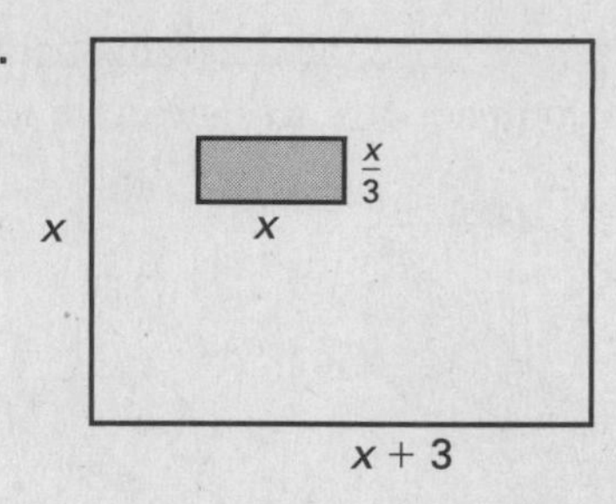

LESSON **9.4**

NAME ______________________ DATE ____________

Practice C

For use with pages 554–560

If possible, simplify the rational expression.

1. $\dfrac{3x^2 - 5x - 2}{x^2 - 4}$

2. $\dfrac{2x + 6}{x^2 - 6x + 9}$

3. $\dfrac{x^2 - 25}{x^3 - 125}$

Multiply the rational expressions. Simplify the result.

4. $(x + 5) \cdot \dfrac{x^2 - 36}{x^2 + 11x + 30}$

5. $\dfrac{x^2 - 2x}{x^2 + 2x + 1} \cdot \dfrac{x^2 + 4x + 3}{x^2 + 3x}$

6. $\dfrac{3x^2 - 12}{5x - 10} \cdot \dfrac{1}{2x + 4}$

7. $\dfrac{21x^{10}y^5}{5x^2} \cdot \dfrac{x^3}{35y^4}$

Divide the rational expressions. Simplify the result.

8. $(x^2 + 10x - 24) \div \dfrac{x^2 - 144}{3x - 36}$

9. $\dfrac{7x^2 - 21x}{x^2 - 2x - 35} \div \dfrac{x^2}{x - 7}$

10. $\dfrac{x^3 - 8}{64x} \div \dfrac{x^2 - x - 2}{16x^2}$

11. $\dfrac{2x^3 - 12x^2}{x^2 - 4x - 12} \div \dfrac{8x^3 + 24x^2}{x^2 + 9x + 18}$

Perform the indicated operations. Simplify the result.

12. $\dfrac{x^2 - 3x + 2}{x + 2} \cdot \dfrac{3x}{x - 2} \cdot \dfrac{2x + 4}{5x^2 - 5x}$

13. $\dfrac{x^2 - 100}{4x^2} \cdot \dfrac{x^3 - 5x^2 - 50x}{x^4 + 10x^3} \div \dfrac{(x - 10)^2}{5x}$

14. $(x^2 + 7x - 30) \div \dfrac{x^2 + 5x - 24}{x + 2} \cdot \dfrac{x + 2}{x^2 + 3x + 2}$

15. $\dfrac{1}{x^3 + 10x^2} \div \dfrac{x^2 - 9}{x + 3} \cdot \dfrac{x + 10}{x^2 + 7x + 12}$

Swimming Pools **In Exercises 16 and 17, use the following information.**

You are considering buying a swimming pool and have narrowed the choices to two—one that is circular and one that is rectangular. The width of the rectangular pool is three times its depth. Its length is 6 feet more than its width. The circular pool has a diameter that is twice the width of the rectangular pool, and it is 2 feet deeper.

16. Find the ratio of the volume of the circular pool to the volume of the rectangular pool.

17. The volume of the rectangular pool is 2592 cubic feet. How many gallons of water are needed to fill the circular pool if 1 gallon is approximately 0.134 cubic foot?

LESSON 9.4

NAME ______________________ DATE __________

Reteaching with Practice

For use with pages 554–560

GOAL **Multiply and divide rational expressions**

VOCABULARY

A rational expression is in **simplified form** provided its numerator and denominator have no common factors, other than ± 1.

Simplifying Rational Expressions

Let a, b, and c be nonzero real numbers or variable expressions. Then the following property applies:

$$\frac{a\cancel{c}}{b\cancel{c}} = \frac{a}{b}$$ Divide out common factor c.

To divide one rational expression by another, multiply the first expression by the reciprocal of the second expression.

$$\frac{a}{b} \div \frac{c}{d} = \frac{a}{b} \cdot \frac{d}{c} = \frac{ad}{bc} \longleftarrow \text{Simplify } \frac{ad}{bc} \text{ if possible.}$$

EXAMPLE 1 *Simplifying a Rational Expression*

Simplify $\frac{x^2 + 5x + 6}{x^3 + 3x^2}$.

SOLUTION

$$\frac{x^2 + 5x + 6}{x^3 + 3x^2} = \frac{(x + 2)(x + 3)}{x^2(x + 3)}$$ Factor numerator and denominator.

$$= \frac{(x + 2)\cancel{(x + 3)}}{x^2\cancel{(x + 3)}}$$ Divide out common factor.

$$= \frac{x + 2}{x^2}$$ Simplified form

Exercises for Example 1

If possible, simplify the rational expression.

1. $\frac{y^2 - 81}{2y - 18}$ **2.** $\frac{2x - 3}{4x - 6}$ **3.** $\frac{x + 3}{x^2 + 6x + 9}$ **4.** $\frac{y^2 - 7y}{y^2 - 8y + 7}$

Lesson 9.4

NAME ______________________ DATE ____________

Reteaching with Practice

For use with pages 554–560

EXAMPLE 2 Multiplying Rational Expressions Involving Polynomials

Multiply: $\dfrac{x^2 - 2x}{x^2 + 2x + 1} \cdot \dfrac{x^2 + 4x + 3}{x^2 + 3x}$

SOLUTION

$$\frac{x^2 - 2x}{x^2 + 2x + 1} \cdot \frac{x^2 + 4x + 3}{x^2 + 3x} = \frac{x(x-2)}{(x+1)(x+1)} \cdot \frac{(x+3)(x+1)}{x(x+3)}$$ Factor numerators and denominators.

$$= \frac{\cancel{x}(x-2)\cancel{(x+3)}\cancel{(x+1)}}{\cancel{x}(x+1)\cancel{(x+1)}\cancel{(x+3)}}$$ Multiply and divide out common factors.

$$= \frac{x-2}{x+1}$$ Simplified form

Exercises for Example 2

Multiply the rational expressions. Simplify the result.

5. $\dfrac{x^2 + 2x - 3}{x + 2} \cdot \dfrac{x^2 + 2x}{x^2 - 1}$

6. $\dfrac{5x - 20}{5x + 15} \cdot \dfrac{2x + 6}{x - 4}$

7. $\dfrac{12 - x}{3} \cdot \dfrac{3}{x - 12}$

EXAMPLE 3 Dividing Rational Expressions

Divide: $\dfrac{2x^3 - 12x^2}{x^2 - 4x - 12} \div \dfrac{8x^3 + 24x^2}{x^2 + 9x + 18}$

SOLUTION

$$\frac{2x^3 - 12x^2}{x^2 - 4x - 12} \div \frac{8x^3 + 24x^2}{x^2 + 9x + 18} = \frac{2x^3 - 12x^2}{x^2 - 4x - 12} \cdot \frac{x^2 + 9x + 18}{8x^3 + 24x^2}$$ Multiply by reciprocal.

$$= \frac{2x^2(x-6)}{(x+2)(x-6)} \cdot \frac{(x+6)(x+3)}{8x^2(x+3)}$$ Factor.

$$= \frac{\cancel{2x^2}\cancel{(x-6)}(x+6)\cancel{(x+3)}}{\cancel{2} \cdot 4\cancel{x^2}(x+2)\cancel{(x-6)}\cancel{(x+3)}}$$ Multiply and divide out common factors.

$$= \frac{x+6}{4(x+2)}$$ Simplified form

Exercises for Example 3

Divide the rational expressions. Simplify the result.

8. $\dfrac{48x^2}{y} \div \dfrac{36xy^2}{5}$

9. $\dfrac{x^2}{x^2 - 1} \div \dfrac{3x}{x + 1}$

LESSON 9.4

NAME ______________________ DATE ________

Quick Catch-Up for Absent Students

For use with pages 554–561

The items checked below were covered in class on (date missed) ____________

Lesson 9.4: Multiplying and Dividing Rational Expressions

____ **Goal 1:** Multiply and divide rational expressions. (pp. 554–556)

Material Covered:

____ Example 1: Simplifying a Rational Expression

____ Example 2: Multiplying Rational Expressions Involving Monomials

____ Example 3: Multiplying Rational Expressions Involving Polynomials

____ Student Help: Look Back

____ Example 4: Multiplying by a Polynomial

____ Example 5: Dividing Rational Expressions

____ Example 6: Dividing by a Polynomial

____ Example 7: Multiplying and Dividing

Vocabulary:

simplified form of a rational expression, p. 554

____ **Goal 2:** Use rational expressions to model real-life quantities. (p. 557)

Material Covered:

____ Example 8: Writing and Simplifying a Rational Model

Activity 9.4: Operations with Rational Expressions (p. 561)

____ **Goal:** Use a graphing calculator to verify the results of problems involving rational expressions.

____ Student Help: Keystroke Help

____ Other (specify) ____

____

Homework and Additional Learning Support

____ Textbook (specify) pp. 558–560

____

____ Internet: Extra Examples at www.mcdougallittell.com

____ *Reteaching with Practice* worksheet (specify exercises)________________

____ *Personal Student Tutor* for Lesson 9.4

NAME ______________________________ DATE __________

Interdisciplinary Application

For use with pages 554–560

Marching Band

MUSIC Most high schools and some middle schools have a music program. There are many different areas in the music department. They include concert band, orchestra, jazz band or stage band, and marching band.

Marching in a band is more than just playing a half-time show for a football crowd. A marching band can be as serious as participating for the national title at the *Marching Bands of America* competition. Some high school marching band members go on to march in Drum and Bugle Corps. Some marching bands march in famous parades, such as the Rose Bowl parade, the Macy's Day parade, or Mardi Gras.

In Exercises 1–5, use the following information.

A new high school opened seven years ago. The number of members, M, in the high school marching band can be modeled by

$$M = \frac{-1316y - 6387}{3.89y - 100}, 0 \le y \le 7,$$

where y is the number of years since the school opened.

The number of trumpet players, T, in the same marching band can be modeled by

$$T = \frac{-245y - 550}{1.79y - 100}, 0 \le y \le 7,$$

where y is the number of years since the school opened.

1. Write a model to find the percent of trumpet players, T, in the marching band for each year.
2. Find the percent of trumpet players two years after the school opened.
3. Find the year(s) in which the percent of trumpet players was greater than or equal to 12%.
4. The band director assumes that the marching band will continue to grow at the same rate. The director wants you to predict how many members will be in the marching band 10 years after the school opened. Of those members how many are expected to be trumpet players?
5. Using your model from Exercise 1, estimate the percent of trumpet players 10 years after the school opened.

NAME ______________________ DATE __________

Challenge: Skills and Applications

For use with pages 554–560

1. a. Let $f(x) = \dfrac{x^2 - x - 6}{x^2 - 1}$, and $g(x) = \dfrac{x^2 - 6x + 5}{x^2 - 4}$. Simplify the function $h(x) = f(x) \cdot g(x)$ and find the equations of the vertical asymptotes of h.

b. Suppose f and g are rational functions such that, for each function, the numerator and denominator have no common factors, and each numerator and denominator has no repeated factors. Make a conjecture relating the vertical asymptotes of the product $f \cdot g$ to the vertical asymptotes of f and g and their zeros.

In Exercises 2 and 3, perform the indicated operations. Simplify the result.

2. $\dfrac{x^3 + 27}{x^2 + 7x + 12} \div \dfrac{x + 3}{x^2 + 8x + 16} \cdot \dfrac{x^2 - 9}{x^2 - 3x + 9}$

3. $\dfrac{x^2 - 25}{x^3 - 125} \div \left(\dfrac{x^2 - 6x + 5}{x^2 - 5x} \cdot \dfrac{x^2 + 5x}{x - 1}\right)$

4. a. Write out the general formula for the expansion of $(x + a)^3$ in terms of x and a.

b. Use the expansion you found in part (a) to help you simplify each product.

(i) $\dfrac{x^3 + 3x^2 + 3x + 1}{x^2 - 1} \cdot \dfrac{x^2 - 3x + 2}{x^2 + 2x + 1}$

(ii) $\dfrac{x^2 - 4}{x^2 - 2x} \cdot \dfrac{x^2 + 5x + 6}{x^3 + 6x^2 + 12x + 8}$

5. a. Show that, for any integer $n \geq 2$,

$(x - a)(x^{n-1} + ax^{n-2} + \cdots + a^{n-3}x^2 + a^{n-2}x + a^{n-1}) = x^n - a^n$.

b. Use the fact stated in part (a) to help you simplify each product.

(i) $\dfrac{x^2 + 4x - 5}{x^4 - 1} \cdot \dfrac{x^3 + x^2 + x + 1}{x^2 + 10x + 25}$

(ii) $\dfrac{x^2 + 2x + 4}{x^4 - 16} \cdot \dfrac{x^3 + 2x^2 + 4x + 8}{x^3 - 8}$

LESSON

TEACHER'S NAME ____________________ CLASS ______ ROOM ______ DATE ______

Lesson Plan

1-day lesson (See *Pacing the Chapter,* TE pages 530C–530D) For use with pages 562–567

GOALS
1. Add and subtract rational expressions.
2. Simplify complex fractions.

State/Local Objectives ____

____

✓ **Check the items you wish to use for this lesson.**

STARTING OPTIONS

____ Homework Check: TE page 558; Answer Transparencies
____ Warm-Up or Daily Homework Quiz: TE pages 562 and 560, CRB page 65, or Transparencies

TEACHING OPTIONS

____ Lesson Opener (Visual Approach): CRB page 66 or Transparencies
____ Graphing Calculator Activity with Keystrokes: CRB pages 67–68
____ Examples 1–6: SE pages 562–564
____ Extra Examples: TE pages 563–564 or Transparencies; Internet
____ Closure Question: TE page 564
____ Guided Practice Exercises: SE page 565

APPLY/HOMEWORK

Homework Assignment

____ Basic 12–14, 18–21, 26–40 even, 52–54, 57–71 odd
____ Average 12–16, 18–25, 26–40 even, 47, 52–54, 57–71 odd
____ Advanced 12–16, 18–25, 26–46 even, 48–54, 57–71 odd

Reteaching the Lesson

____ Practice Masters: CRB pages 69–71 (Level A, Level B, Level C)
____ Reteaching with Practice: CRB pages 72–73 or Practice Workbook with Examples
____ Personal Student Tutor

Extending the Lesson

____ Cooperative Learning Activity: CRB page 75
____ Applications (Real-Life): CRB page 76
____ Challenge: SE page 567; CRB page 77 or Internet

ASSESSMENT OPTIONS

____ Checkpoint Exercises: TE pages 563–564 or Transparencies
____ Daily Homework Quiz (9.5): TE page 567, CRB page 80, or Transparencies
____ Standardized Test Practice: SE page 567; TE page 567; STP Workbook; Transparencies

Notes ____

____

____

LESSON 9.5

TEACHER'S NAME ____________ CLASS ______ ROOM ______ DATE ______

Lesson Plan for Block Scheduling

Half-day lesson (See *Pacing the Chapter,* TE pages 530C–530D) **For use with pages 562–567**

GOALS
1. **Add and subtract rational expressions.**
2. **Simplify complex fractions.**

State/Local Objectives ____________

CHAPTER PACING GUIDE	
Day	**Lesson**
1	9.1 (all); 9.2(all)
2	9.3 (all)
3	9.4 (all)
4	**9.5 (all)**; 9.6(all)
5	Review/Assess Ch. 9

✓ Check the items you wish to use for this lesson.

STARTING OPTIONS

____ Homework Check: TE page 558; Answer Transparencies
____ Warm-Up or Daily Homework Quiz: TE pages 562 and 560, CRB page 65, or Transparencies

TEACHING OPTIONS

____ Lesson Opener (Visual Approach): CRB page 66 or Transparencies
____ Graphing Calculator Activity with Keystrokes: CRB pages 67–68
____ Examples 1–6: SE pages 562–564
____ Extra Examples: TE pages 563–564 or Transparencies; Internet
____ Closure Question: TE page 564
____ Guided Practice Exercises: SE page 565

APPLY/HOMEWORK

Homework Assignment (See also the assignment for Lesson 9.6.)

____ Block Schedule: 12–16, 18–25, 26–40 even, 47, 52–54, 57–71 odd

Reteaching the Lesson

____ Practice Masters: CRB pages 69–71 (Level A, Level B, Level C)
____ Reteaching with Practice: CRB pages 72–73 or Practice Workbook with Examples
____ Personal Student Tutor

Extending the Lesson

____ Cooperative Learning Activity: CRB page 75
____ Applications (Real-Life): CRB page 76
____ Challenge: SE page 567; CRB page 77 or Internet

ASSESSMENT OPTIONS

____ Checkpoint Exercises: TE pages 563–564 or Transparencies
____ Daily Homework Quiz (9.5): TE page 567, CRB page 80, or Transparencies
____ Standardized Test Practice: SE page 567; TE page 567; STP Workbook; Transparencies

Notes ____________

LESSON 9.5

NAME ______________________ DATE ________

Available as a transparency

Warm-Up Exercises

For use before Lesson 9.5, pages 562–567

Simplify.

1. $\frac{3}{5} + \frac{1}{5}$

2. $\frac{3}{4} + \frac{1}{2}$

3. $\frac{2}{3} - \frac{1}{2}$

4. $\dfrac{\frac{1}{2}}{\frac{3}{4}}$

5. $\dfrac{1}{\frac{1}{2} + \frac{1}{3}}$

Daily Homework Quiz

For use after Lesson 9.4, pages 554–561

1. If possible, simplify the rational expression.

   a. $\dfrac{x^2 - x - 6}{x^2 + 6x + 8}$

   b. $\dfrac{x^2 - 16}{x^2 + 8x + 16}$

2. Perform the indicated operation. Simplify the result.

   a. $\dfrac{x^2 - 1}{12x^2 + 24x} \cdot \dfrac{4}{x^2 + x}$

   b. $\dfrac{x^2 + 5x + 6}{x^3 - x^2} \div \dfrac{x + 3}{x^2}$

   c. $\dfrac{6x^2 + 7x + 1}{7x + 49} \div \dfrac{2x + 2}{2x + 14}$

   d. $\dfrac{x + 1}{6x - 3} \cdot \dfrac{3x^2}{x^2 + x} \div \dfrac{x^2}{2x^2 - x}$

LESSON 9.5

NAME ______________________ DATE __________

Available as a transparency

Visual Approach Lesson Opener

For use with pages 562–567

The process of adding rational expressions is similar to methods you have previously learned for adding fractions.

For example, to simplify $\frac{2}{3} + \frac{5}{7}$, you would first find the lowest common denominator. Because 3 and 7 have no common factors, the LCD is $3 \cdot 7 = 21$. You would then rewrite each expression using the LCD. Add the numerators, and simplify if possible.

$$\frac{2}{3} + \frac{5}{7} = \left(\frac{2}{3} \cdot \frac{7}{7}\right) + \left(\frac{5}{7} \cdot \frac{3}{3}\right) = \frac{14}{21} + \frac{15}{21} = \frac{29}{21}$$

Fill in the missing expressions to simplify $\frac{2}{x-1} + \frac{5}{x+3}$.

Use the fraction problem above as a guide.

1. Find the LCD. Because __________ and __________ have no common factors, the LCD is (__________)(__________).

2. Rewrite each expression using the LCD. Then add the numerators, and simplify if possible.

$$\frac{2}{x-1} + \frac{5}{x+3} = \frac{2}{x-1} \cdot \frac{\blacksquare}{\blacksquare} + \frac{5}{x+3} \cdot \frac{\blacksquare}{\blacksquare}$$

$$= \frac{\blacksquare}{(\blacksquare)(\blacksquare)} + \frac{\blacksquare}{(\blacksquare)(\blacksquare)}$$

$$= \frac{\blacksquare}{(\blacksquare)(\blacksquare)}$$

LESSON 9.5

NAME ______________________________ DATE ________

Graphing Calculator Activity

For use with pages 562–567

GOAL **To use a graphing calculator to verify the algebra involved in the addition of fractions with like denominators and in the simplification of complex fractions**

As with numerical fractions, the procedure used to add two rational expressions depends upon whether the expressions have *like* or *unlike* denominators. To add two rational expressions with like denominators, simply add their numerators and place the result over the common denominator.

A *complex fraction* is a fraction that contains a fraction in its numerator or denominator.

Activities

1. Use algebra to see that $\frac{2}{3x} + \frac{5}{3x} = \frac{7}{3x}$.

2. Use a graphing calculator to verify the equation in Step 1. Let y_1 be $\frac{2}{3x} + \frac{5}{3x}$ and let y_2 be $\frac{7}{3x}$. Graph y_1 and y_2 on the same screen. If the graphs coincide, then the expressions are equivalent.

3. Use algebra to simplify $\dfrac{2 + \frac{3}{x}}{\frac{5}{x} + 3}$ by multiplying both the numerator and the denominator by x. The answer should be $\frac{2x + 3}{5 + 3x}$.

4. Use a graphing calculator to verify the equation in Step 3. Let y_1 be $\dfrac{2 + \frac{3}{x}}{\frac{5}{x} + 3}$ and let y_2 be $\frac{2x + 3}{5 + 3x}$. Graph y_1 and y_2 on the same screen. If the graphs coincide, then the expressions are equivalent.

Exercises

1. By finding a common denominator and creating equivalent fractions, use algebra to show that $\frac{2}{x + 2} + \frac{3}{x - 5}$ is equivalent to $\frac{5x - 4}{(x + 2)(x - 5)}$.
2. Use a graphing calculator to show that the two expressions in Exercise 1 are equivalent.
3. Use algebra to simplify $\dfrac{\frac{2}{x} + 3}{\frac{1}{x + 2} + 5}$ by multiplying both the numerator and the denominator by $x(x + 2)$.
4. Use a graphing calculator to verify your work in Exercise 3.

NAME ______________________ DATE __________

Graphing Calculator Activity

For use with pages 562–567

TI-82

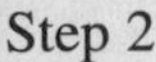

Step 2

Y= 2 ÷ (3 X,T,θ) + 5 ÷ (
3 X,T,θ) ENTER
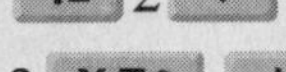
7 ÷ (3 X,T,θ) ENTER
ZOOM 6

Step 4

Y= (2 + 3 ÷ X,T,θ) ÷ (
5 ÷ X,T,θ + 3) ENTER
(2 X,T,θ + 3) ÷ (5 + 3 X,T,θ
) ENTER GRAPH

TI-83

Step 2

Y= 2 ÷ (3 X,T,θ,n) + 5 ÷ (
3 X,T,θ,n) ENTER
7 ÷ (3 X,T,θ,n) ENTER
ZOOM 6

Step 4

Y= (2 + 3 ÷ X,T,θ,n) ÷ (
5 ÷ X,T,θ,n + 3) ENTER
(2 X,T,θ,n + 3) ÷ (5 + 3
X,T,θ,n) ENTER

SHARP EL-9600c

Step 2

Y= 2 ÷ (3 X/θ/T/n) + 5 ÷ (
3 X/θ/T/n) ENTER
7 ÷ (3 X/θ/T/n) ENTER
ZOOM [A]5

Step 4

Y= (2 + 3 ÷ X/θ/T/n) ÷ (
5 ÷ X/θ/T/n + 3) ENTER
(2 X/θ/T/n + 3) ÷ (5 + 3
X/θ/T/n) ENTER GRAPH

CASIO CFX-9850GA PLUS

From the main menu, select TABLE.

1. Enter functions

2 ÷ (3 X,θ,T) + 5 ÷ (3
X,θ,T) EXE
7 ÷ (3 X,θ,T) EXE
SHIFT F3 F3 EXIT F6

Step 4

(2 + 3 ÷ X,θ,T) ÷ (5
÷ X,θ,T + 3) EXE
(2 X,θ,T + 3) ÷ (5 + 3 X,θ,T
) EXE F6

LESSON 9.5

NAME ______________________________ DATE __________

Practice A

For use with pages 562–567

Perform the indicated operation and simplify.

1. $\frac{7}{4x} + \frac{9}{4x}$

2. $\frac{5}{x+1} + \frac{x}{x+1}$

3. $\frac{2x}{x-3} - \frac{1}{x-3}$

Find the least common denominator.

4. $\frac{5}{x-1}, \frac{6}{x^2-1}$

5. $\frac{3}{x+4}, \frac{x}{x^2-16}$

6. $\frac{x}{x^2+x-2}, \frac{1}{x+2}$

7. $\frac{5}{2(x+1)}, \frac{1}{2x}, \frac{3}{2(x+1)^2}$

Perform the indicated operation(s) and simplify.

8. $\frac{5}{x} + \frac{2}{3x^2}$

9. $\frac{1}{2} + \frac{2}{x} - \frac{3}{x^2}$

10. $\frac{3}{x+5} - \frac{4}{x+1}$

11. $6 - \frac{5}{x+3}$

12. $\frac{4x}{x^2-4} - \frac{3}{x+2}$

13. $\frac{x}{x-1} + \frac{3x}{x^2-1}$

Simplify the complex fraction.

14. $\frac{\left(2 - \frac{1}{x}\right)}{x}$

15. $\frac{\left(1 + \frac{1}{x}\right)}{\left(1 - \frac{1}{x}\right)}$

16. $\frac{\left(\frac{6}{x-1} - 3\right)}{\left(\frac{3}{x}\right)}$

Electrical Resistors **In Exercises 17 and 18, use the following information.**

When two resistors with resistances R_1 and R_2 are connected in parallel, the total resistance R is given by $R = \frac{1}{\left(\frac{1}{R_1} + \frac{1}{R_2}\right)}$.

17. Simplify this complex fraction.

18. Find the total resistance (in ohms) of a 4 ohm resistor and a 2 ohm resistor that are connected in parallel.

LESSON 9.5

Lesson 9.5

NAME ______________________________ DATE __________

Practice B

For use with pages 562–567

Find the least common denominator.

1. $\frac{5}{2x+1}, \frac{6}{4x^2-1}$

2. $\frac{3}{x+4}, \frac{x+2}{4}$

3. $\frac{4}{x^2-1}, \frac{5}{x(x+1)}$

Perform the indicated operation(s) and simplify.

4. $\frac{7}{x-2} + \frac{x}{x-2}$

5. $\frac{x}{x^2+x-2} + \frac{1}{x+2}$

6. $\frac{x}{x^2-x-30} - \frac{1}{x+5}$

7. $\frac{4}{x} - \frac{2}{x^2} + \frac{4}{x+3}$

8. $\frac{x+2}{x-1} - \frac{2}{x+6} - \frac{14}{x^2+5x-6}$

9. $\frac{x}{x^2-9} + \frac{3}{x(x-3)}$

10. $4 - \frac{5}{x+3}$

11. $\frac{1}{3} + \frac{3}{x} - \frac{4}{x^2}$

Simplify the complex fraction.

12. $\dfrac{\left(\frac{1}{x} + \frac{1}{2x+1}\right)}{\left(\frac{4x}{2x+1}\right)}$

13. $\dfrac{\left(\frac{1}{3x} - \frac{4}{x+2}\right)}{\left(\frac{x}{x+2} + \frac{1}{x}\right)}$

14. $\dfrac{\left(\frac{2}{4x+12}\right)}{\left(\frac{4}{2x+6} + \frac{1}{x+3}\right)}$

Doctors **In Exercises 15 and 16, use the following information.**

Over a twenty-year period the number of doctors of medicine M (in thousands) in the United States can be approximated by $M = \frac{28{,}390 + 693t}{85 - t}$ where $t = 0$ represents 1980. The number of doctors of osteopathy B (in thousands) can be approximated by $B = \frac{776 - 12t}{55 - 2t}$.

15. Write an expression for the total number I of doctors of medicine (MD) and doctors of osteopathy (DO). Simplify the result.

16. How many MDs and DOs did the United States have in 1995?

NAME ______________________________ DATE ____________

Practice C

For use with pages 562–567

Find the least common denominator.

1. $\dfrac{3}{x+4}, \dfrac{x}{x^2-16}, \dfrac{x+2}{4}$

2. $\dfrac{13}{x^2-2x+1}, \dfrac{4}{x^2-1}, \dfrac{5}{x(x+1)}$

3. $\dfrac{7}{x-6}, \dfrac{5x}{x(x-2)}, \dfrac{3}{x^2-8x+12}$

Perform the indicated operation(s) and simplify.

4. $\dfrac{2x-1}{x^2-x-2} - \dfrac{1}{x-2}$

5. $\dfrac{5}{3x-12} + \dfrac{3x+1}{x^2-x-12} - \dfrac{2}{3}$

6. $\dfrac{2}{x} + \dfrac{3x+1}{x^2} - \dfrac{x-2}{x^3}$

7. $\dfrac{3x}{x+2} + \dfrac{5x}{x-2} - \dfrac{40}{x^2-4}$

8. $\dfrac{2x}{x+2} - \dfrac{8}{x^2+2x} + \dfrac{3}{x}$

9. $\dfrac{2x+1}{x^2+4x+4} - \dfrac{6x}{x^2-4} + \dfrac{3}{x-2}$

10. $\dfrac{2x+5}{x^2+6x+9} + \dfrac{x}{x^2-9} + \dfrac{1}{x-3}$

11. $\dfrac{5}{2(x+1)} - \dfrac{1}{2x} - \dfrac{3}{2(x+1)^2}$

Simplify the complex fraction.

12. $\dfrac{\left(\dfrac{1}{x+9} + \dfrac{1}{5}\right)}{\left(\dfrac{2}{x^2+10x+9}\right)}$

13. $\dfrac{\left(\dfrac{4}{x^2-25} + \dfrac{2}{x+5}\right)}{\left(\dfrac{1}{x+5} + \dfrac{1}{x-5}\right)}$

14. $\dfrac{\left(\dfrac{x}{x-4} - \dfrac{1}{4}\right)}{\left(\dfrac{9}{4x} + \dfrac{x^2}{x-4}\right)}$

Electronics Pattern **In Exercises 15 and 16, use the following information.**

The total resistance R_t (in ohms) of three resistors in a parallel circuit is given by the formula $R_t = \dfrac{1}{\dfrac{1}{R_1} + \dfrac{1}{R_2} + \dfrac{1}{R_3}}$, which can be simplified to

$$R_t = \frac{R_1R_2R_3}{R_1R_2 + R_1R_3 + R_2R_3}.$$

15. Simplify the similar formula for four resistors in a parallel circuit given by the formula $R_t = \dfrac{1}{\dfrac{1}{R_1} + \dfrac{1}{R_2} + \dfrac{1}{R_3} + \dfrac{1}{R_4}}$.

16. Following the pattern (without algebraically simplifying the complex fraction), write the simplified formula for the total resistance R_t (in ohms) of five resistors in a parallel circuit.

LESSON 9.5

NAME ______________________ DATE __________

Reteaching with Practice

For use with pages 562–567

GOAL Add and subtract rational expressions and simplify complex fractions

VOCABULARY

A **complex fraction** is a fraction that contains a fraction in its numerator or denominator.

EXAMPLE 1 *Adding with Unlike Denominators*

Add: $\frac{5}{2x} + \frac{2}{3x^2}$.

SOLUTION

Begin by finding the least common denominator of $\frac{5}{2x}$ and $\frac{2}{3x^2}$. Notice that the denominators are already written as factors, and the LCD is $6x^2$.

$$\frac{5}{2x} + \frac{2}{3x^2} = \frac{5(3x)}{2x(3x)} + \frac{2(2)}{3x^2(2)}$$ Rewrite fractions with LCD.

$$= \frac{15x}{6x^2} + \frac{4}{6x^2} = \frac{15x + 4}{6x^2}$$ Simplify and add numerators.

Exercises for Example 1

Perform the indicated operation and simplify.

1. $\frac{1}{2} + \frac{3}{x^2}$

2. $\frac{3}{2x} + \frac{x}{2x^2 + 6x}$

3. $\frac{3}{x + 5} + \frac{4}{x + 1}$

4. $\frac{x}{x - 1} + \frac{3x}{x^2 - 1}$

EXAMPLE 2 *Subtracting with Unlike Denominators*

Subtract: $\frac{3x + 1}{x^2 - x - 12} - \frac{5}{3x - 12}$.

SOLUTION

$$\frac{3x + 1}{x^2 - x - 12} - \frac{5}{3x - 12} = \frac{3x + 1}{(x - 4)(x + 3)} - \frac{5}{3(x - 4)}$$ Factor denominators.

$$= \frac{(3x + 1)(3)}{(x - 4)(x + 3)(3)} - \frac{5(x + 3)}{3(x - 4)(x + 3)}$$ Rewrite fractions with LCD.

$$= \frac{9x + 3 - 5(x + 3)}{3(x - 4)(x + 3)}$$ Subtract numerators.

$$= \frac{9x + 3 - 5x - 15}{3(x - 4)(x + 3)}$$ Distribute.

$$= \frac{4x - 12}{3(x - 4)(x + 3)}$$ Simplify.

NAME ______________________ DATE __________

Reteaching with Practice

For use with pages 562–567

Exercises for Example 2

Perform the indicated operation and simplify.

5. $\dfrac{2x}{x+2} - \dfrac{8}{x^2+2x}$

6. $\dfrac{5x}{x^2-4} - \dfrac{7}{x-2}$

7. $\dfrac{3x+1}{x^2} - \dfrac{x-2}{x^3}$

8. $\dfrac{x}{x+3} - \dfrac{6}{x+2}$

EXAMPLE 3 Simplifying a Complex Fraction

Simplify: $\dfrac{\dfrac{6}{x-1} - 3}{\dfrac{3}{x}}$.

SOLUTION

$$\frac{\dfrac{6}{x-1} - 3}{\dfrac{3}{x}} = \frac{\dfrac{6}{x-1} - \dfrac{3(x-1)}{x-1}}{\dfrac{3}{x}}$$ Rewrite fractions in numerator with LCD.

$$= \frac{\dfrac{3(3-x)}{x-1}}{\dfrac{3}{x}}$$ Subtract fractions in numerator.

$$= \frac{3(3-x)}{x-1} \cdot \frac{x}{3}$$ Multiply by reciprocal.

$$= \frac{\cancel{3}(3-x)}{x-1} \cdot \frac{x}{\cancel{3}}$$ Divide out common factor.

$$= \frac{x(3-x)}{x-1}$$ Write in simplified form.

Exercises for Example 3

Simplify the complex fraction.

9. $\dfrac{\dfrac{x^2}{x^2-1}}{\dfrac{3x}{x+1}}$

10. $\dfrac{2 - \dfrac{1}{x}}{x}$

11. $\dfrac{1 + \dfrac{1}{x}}{1 - \dfrac{1}{x}}$

LESSON 9.5

NAME ______________________________ DATE __________

Quick Catch-Up for Absent Students

For use with pages 562–567

The items checked below were covered in class on (date missed) __________

Lesson 9.5: Addition, Subtraction, and Complex Fractions

____ **Goal 1:** Add and subtract rational expressions. (pp. 562–563)

Material Covered:

____ Example 1: Adding and Subtracting with Like Denominators

____ Student Help: Skills Review

____ Example 2: Adding with Unlike Denominators

____ Student Help: Look Back

____ Example 3: Subtracting with Unlike Denominators

____ Example 4: Adding Rational Models

____ **Goal 2:** Simplify complex fractions. (p. 564)

Material Covered:

____ Example 5: Simplifying a Complex Fraction

____ Example 6: Simplifying a Complex Fraction

Vocabulary:

complex fraction, p. 564

____ Other (specify) ______________________________

____

Homework and Additional Learning Support

____ Textbook (specify) pp. 565–567

____

____ Internet: Extra Examples at www.mcdougallittell.com

____ *Reteaching with Practice* worksheet (specify exercises) __________

____ *Personal Student Tutor* for Lesson 9.5

LESSON 9.5

NAME ______________________________ DATE ________

Cooperative Learning Activity

For use with pages 562–567

GOAL **To use a complex fraction to find a monthly loan payment**

Materials: calculator

Background

Monthly payments for items that are financed such as a car, house, or stereo system, will vary with the rate of interest, the amount of time, the loan, and the length of the loan.

Instructions

1. Use the formula $M = P\left[\dfrac{\frac{r}{12}}{1 - \left(\dfrac{1}{1 + \frac{r}{12}}\right)^{12t}}\right]$

   where P is the principle, r is the annual interest rate in decimal form, and t is the length of the loan in years.

2. Find the monthly payment when $P = 16{,}000$, $r = 0.15$, and $t = 1$.

3. Find the monthly payment for financing a $12,500 car for 3 years using interest rates of **a.** 8%, **b.** 12%, and **c.** 18%.

Analyzing the Results

1. For each interest rate in Step 3, find how much more you would pay for the car.

NAME ______________________ DATE __________

Real-Life Application: When Will I Ever Use This?

For use with pages 562–567

Commercial Banks

Commercial banks are the most common banks in the United States. These banks offer many services to individual customers but they primarily work for the needs of businesses.

A bank has offices called branches. Most people say they have to go to the bank, but actually they go to the branch office. A bank is an institution or a company. The building where an office is located is a branch office.

In Exercises 1–4, use the following information.

For 1990 through 1992, the number of banks in the United States can be modeled by $K = \frac{1{,}228{,}230 + 36{,}590t^2}{100 + 4.66t^2}$, where K is the number of banks and t is the year, with $t = 0$ corresponding to 1990.

For 1990 through 1992, the number of branches in the United States can be modeled by $R = \frac{505{,}430 + 9661t^2}{10 + 0.128t^2}$, where R is the number of branches and t is the year, with $t = 0$ corresponding to 1990.

1. Find the total number of commercial banks in the United States in 1992.
2. In what year did the total number of commercial banks grow to more than 66,000?
3. To find the average number of branches per bank, divide the number of branches by the number of banks. Write an equation that represents the average number of branches per bank. Simplify the result.
4. Find the average number of branches per bank in 1996.

LESSON

9.5

NAME ______________________________ DATE __________

Challenge: Skills and Applications

For use with pages 562–567

1. a. Write a function whose value for each number x is the sum of x and its reciprocal. Find the turning points of this function by graphing.

b. For what positive number x is the sum of x and its reciprocal a minimum? Does this quantity have a maximum value? Explain why or why not.

c. For what positive number x is the sum of x and *the square* of its reciprocal a minimum? (If you don't recognize this number, try cubing it.)

In Exercises 2–7, simplify the expression.

2. $\dfrac{(x + a)(x^{-1} - a^{-1})}{(x - a)(x^{-1} + a^{-1})}$

3. $\dfrac{\frac{1}{a} + \frac{1}{x}}{\frac{x}{a} - \frac{a}{x}}$

4. $\dfrac{b^{-2} - x^{-2}}{b^{-1} + x^{-1}}$

5. $\left(\dfrac{x}{b} - \dfrac{b}{x}\right) \div \left(1 - \dfrac{b}{x}\right)$

6. $\dfrac{x}{x^2 - 4} + \dfrac{2}{x^2 - 2x} - \dfrac{x + 1}{x^2 + 2x}$

7. $\left(x - a + \dfrac{2a^2}{x + a}\right)\left(\dfrac{1}{x^2} - \dfrac{1}{a^2}\right) \div \left(x^3 - \dfrac{ax^3 + a^4}{x + a}\right)$

8. a. Show, by finding A and B in terms of p and q, that if p and q are given, with $p \neq q$, then there are unique values A and B that make the following equation true.

$$\frac{1}{(x - p)(x - q)} = \frac{A}{x - p} + \frac{B}{x - q}$$

This is called the *partial-fraction decomposition* of the left-hand side. (*Hint*: After the right side is simplified, the numerators must be equal as polynomials.)

b. Use your answer to part (a) to find the partial-fraction decomposition of $\dfrac{1}{x^2 - x - 6}$.

TEACHER'S NAME ______________ CLASS ______ ROOM ______ DATE ______

Lesson Plan

1-day lesson (See *Pacing the Chapter,* TE pages 530C–530D) **For use with pages 568–574**

1. **Solve rational equations.**
2. **Use rational equations to solve real-life problems.**

State/Local Objectives ______________________________________

____

✓ Check the items you wish to use for this lesson.

STARTING OPTIONS

____ Homework Check: TE page 565; Answer Transparencies
____ Warm-Up or Daily Homework Quiz: TE pages 568 and 567, CRB page 80, or Transparencies

TEACHING OPTIONS

____ Motivating the Lesson: TE page 569
____ Lesson Opener (Graphing Calculator): CRB page 81 or Transparencies
____ Examples 1–6: SE pages 568–570
____ Extra Examples: TE pages 569–570 or Transparencies; Internet
____ Closure Question: TE page 570
____ Guided Practice Exercises: SE page 571

APPLY/HOMEWORK

Homework Assignment

____ Basic 15–18, 21–27, 33–38, 42–46 even, 58–61, 63–75 odd; Quiz 2: 1–9
____ Average 15–20, 22–32 even, 33–38, 42–46 even, 51–53, 58–61, 63–75 odd; Quiz 2: 1–9
____ Advanced 15–20, 22–50 even, 51–53, 57–62, 63–75 odd; Quiz 2: 1–9

Reteaching the Lesson

____ Practice Masters: CRB pages 82–84 (Level A, Level B, Level C)
____ Reteaching with Practice: CRB pages 85–86 or Practice Workbook with Examples
____ Personal Student Tutor

Extending the Lesson

____ Applications (Interdisciplinary): CRB page 88
____ Math & History: SE page 574; CRB page 89; Internet
____ Challenge: SE page 573; CRB page 90 or Internet

ASSESSMENT OPTIONS

____ Checkpoint Exercises: TE pages 569–570 or Transparencies
____ Daily Homework Quiz (9.6): TE page 573 or Transparencies
____ Standardized Test Practice: SE page 573; TE page 573; STP Workbook; Transparencies
____ Quiz (9.4–9.6): SE page 574; CRB page 91

Notes ______________________________________

____

____

LESSON 9.6

TEACHER'S NAME ______________ CLASS ______ ROOM ______ DATE ______

Lesson Plan for Block Scheduling

Half-day lesson (See *Pacing the Chapter,* TE pages 530C–530D) **For use with pages 568–574**

GOALS
1. **Solve rational equations.**
2. **Use rational equations to solve real-life problems.**

State/Local Objectives ______________________________

CHAPTER PACING GUIDE	
Day	**Lesson**
1	9.1 (all); 9.2(all)
2	9.3 (all)
3	9.4 (all)
4	9.5 (all); **9.6(all)**
5	Review/Assess Ch. 9

✓ Check the items you wish to use for this lesson.

STARTING OPTIONS

____ Homework Check: TE page 565; Answer Transparencies
____ Warm-Up or Daily Homework Quiz: TE pages 568 and 567, CRB page 80, or Transparencies

TEACHING OPTIONS

____ Motivating the Lesson: TE page 569
____ Lesson Opener (Graphing Calculator): CRB page 81 or Transparencies
____ Examples 1–6: SE pages 568–570
____ Extra Examples: TE pages 569–570 or Transparencies; Internet
____ Closure Question: TE page 570
____ Guided Practice Exercises: SE page 571

APPLY/HOMEWORK

Homework Assignment (See also the assignment for Lesson 9.5.)
____ Block Schedule: 15–20, 22–32 even, 33–38, 42–46 even, 51–53, 58–61, 63–75 odd; Quiz 2: 1–9

Reteaching the Lesson
____ Practice Masters: CRB pages 82–84 (Level A, Level B, Level C)
____ Reteaching with Practice: CRB pages 85–86 or Practice Workbook with Examples
____ Personal Student Tutor

Extending the Lesson
____ Applications (Interdisciplinary): CRB page 88
____ Math & History: SE page 574; CRB page 89; Internet
____ Challenge: SE page 573; CRB page 90 or Internet

ASSESSMENT OPTIONS

____ Checkpoint Exercises: TE pages 569–570 or Transparencies
____ Daily Homework Quiz (9.6): TE page 573 or Transparencies
____ Standardized Test Practice: SE page 573; TE page 573; STP Workbook; Transparencies
____ Quiz (9.4–9.6): SE page 574; CRB page 91

Notes ______________________________

LESSON 9.6

NAME ______________________ DATE ________

Available as a transparency

WARM-UP EXERCISES

For use before Lesson 9.6, pages 568–574

Solve the equation.

1. $(x - 1)(x + 2) = 0$
2. $6x = -18$
3. $4 - 2x = 12$
4. $x^2 - 16 = 9$
5. $x^3 - 25x = 0$

Lesson 9.6

DAILY HOMEWORK QUIZ

For use after Lesson 9.5, pages 562–567

Find the least common denominator.

1. $\dfrac{7}{x - 3}, \dfrac{x}{2(x + 3)}$

2. $\dfrac{-5}{6x^2}, \dfrac{x^2}{2(x + 3)}$

Perform the indicated operation and simplify.

3. $\dfrac{x}{x - 4} + \dfrac{5}{x - 4}$

4. $\dfrac{5}{2x^2} - \dfrac{3}{4x}$

5. $\dfrac{x + 1}{x^2 - x - 6} + \dfrac{5}{x + 2}$

Simplify the complex fraction.

6. $\dfrac{\dfrac{2x}{3} - 2}{1 + \dfrac{12}{x}}$

LESSON **9.6**

NAME ______________________ DATE __________

Available as a transparency

Graphing Calculator Lesson Opener

For use with pages 568–574

In Lesson 9.6, you will learn algebraic methods for solving equations that contain rational expressions. You can obtain approximate solutions to these equations using a graphing calculator.

For example, to solve $\frac{2}{x-2} = 2 - \frac{1}{x}$,

graph $y_1 = \frac{2}{x-2}$ and $y_2 = 2 - \frac{1}{x}$.

Use the *Intersect* feature. The solutions are $x \approx 0.31$ and $x \approx 3.19$.

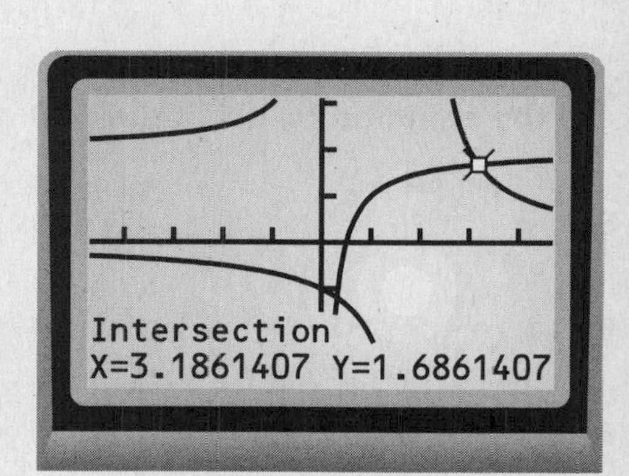

Lesson 9.6

The number of solutions may vary. You may need to change the viewing window to find all solutions.

Use a graphing calculator to solve the equation. If necessary, round to the nearest hundredth.

1. $\frac{3-x}{4} = \frac{x-2}{x+4}$

2. $\frac{2}{x} + \frac{x}{2} = 8$

3. $\frac{x}{x-2} = 3 - \frac{5}{x-2}$

4. $\frac{x}{x+3} = 2 - \frac{3}{x-1}$

5. $\frac{5x}{x-2} = \frac{3x}{x+2}$

6. $\frac{x+2}{x-5} + 6 = \frac{4}{x}$

7. $\frac{3}{x} + \frac{4}{5} = \frac{2}{x}$

8. $\frac{2}{x^2-5} = 5 - \frac{4}{x^2+2x-3}$

LESSON
9.6

NAME ______________________ DATE ________

Practice A

For use with pages 568–574

Determine whether the given x-value is a solution of the equation.

1. $\frac{2}{x-3} = \frac{3}{x+1}, x = -1$

2. $\frac{7}{x+3} = \frac{x}{4}, x = 4$

3. $\frac{x}{x-5} + 4 = \frac{1}{x+3}, x = 4$

4. $\frac{3x-1}{x-2} + 3 = \frac{x}{x-2}, x = -2$

Solve the equation by using the LCD. Check each solution.

5. $\frac{3}{x} - \frac{2}{x+1} = \frac{4}{x}$

6. $\frac{x}{x-4} + 1 = \frac{4}{x-4}$

7. $\frac{15}{x} - 4 = \frac{6}{x} + 3$

8. $\frac{4}{x} - \frac{1}{x+2} = \frac{2}{x}$

9. $\frac{2x}{x+3} + 5 = \frac{3}{x+3}$

10. $\frac{1}{x+2} + \frac{1}{x+2} = \frac{4}{x^2-4}$

Solve the equation by cross multiplying. Check each solution.

11. $\frac{2x-3}{x+3} = \frac{3x}{x+4}$

12. $\frac{x}{2x+1} = \frac{5}{4-x}$

13. $\frac{x}{x-3} = \frac{6}{x-3}$

14. $-\frac{2}{x-1} = \frac{x-8}{x+1}$

15. $\frac{7}{x+3} = \frac{x}{4}$

16. $\frac{x}{x^2-10} = \frac{3}{2x+1}$

Solve the equation using any method. Check each solution.

17. $\frac{3}{x-1} - 6 = \frac{5x}{x-1}$

18. $\frac{5x}{x-1} - 2 = \frac{14}{x^2-1}$

19. $\frac{5x-7}{x-2} = \frac{8}{x-2}$

20. $\frac{1}{x-5} + \frac{1}{x+5} = \frac{x+3}{x^2-25}$

21. $\frac{2x-4}{x-4} = \frac{4}{x-4}$

22. $\frac{1}{x-2} + \frac{1}{x+3} = \frac{5}{x^2+x-6}$

23. ***Population Density*** The population density in a large city is related to the distance from the center of the city. It can be modeled by

$$D = \frac{5000x}{x^2 + 36}$$

where D is the population density (in people per square mile) and x is the distance (in miles) from the center of the city. Find the areas where the population density is 400 people per square mile.

LESSON **9.6**

NAME ______________________________ DATE __________

Practice B

For use with pages 568–574

Determine whether the given *x*-value is a solution of the equation.

1. $\frac{1}{x-3} + \frac{1}{x+3} = \frac{10}{x^2-9}, x = 5$

2. $\frac{x}{x-4} + 1 = \frac{4}{x-4}, x = 4$

Solve the equation by using the LCD. Check each solution.

3. $\frac{3x}{x-2} = 1 + \frac{6}{x-2}$

4. $\frac{3x}{x-2} + \frac{1}{x+2} = -\frac{4}{x^2-4}$

5. $\frac{2}{2x+5} + \frac{3}{2x-5} = \frac{5x+5}{4x^2-25}$

6. $\frac{5}{2x+3} + \frac{4}{2x-3} = \frac{14x+3}{4x^2-9}$

7. $-\frac{15}{x} - 4 = \frac{6}{x} + 3$

8. $\frac{3x-1}{x-2} + 3 = \frac{x}{x-2}$

Solve the equation by cross multiplying. Check each solution.

9. $\frac{x+1}{x+3} = 2$

10. $\frac{2}{x-3} = \frac{3}{x+1}$

11. $\frac{7}{x+3} = \frac{x}{4}$

12. $\frac{6+5x}{3x} = \frac{7}{x}$

13. $\frac{x}{x^2-8} = \frac{2}{x}$

14. $\frac{2x}{5} = \frac{x^2-5x}{5x}$

Solve the equation using any method. Check each solution.

15. $\frac{5x}{x-2} = 7 + \frac{10}{x-2}$

16. $\frac{2x}{4-x} = \frac{x^2}{x-4}$

17. $\frac{3x}{x+1} = \frac{12}{x^2-1} + 2$

18. $\frac{6}{x} - \frac{7x}{5} = \frac{x}{10}$

19. $\frac{3}{x} + 12 = 2 + \frac{4}{3x}$

20. $\frac{x^2+2x+2}{x-1} = \frac{2x+3}{x-1}$

21. ***Average Cost*** A greeting card manufacturer can produce a dozen cards for \$6.50. If the initial investment by the company was \$60,000, how many dozen cards must be produced before the average cost per dozen falls to S11.50?

22. ***Brakes*** The braking distance of a car can be modeled by $d = s + \frac{s^2}{20}$ where d is the distance (in feet) that the car travels before coming to a stop, and s is the speed at which the car is traveling (in miles per hour). Find the speed that results in a braking distance of 75 feet.

LESSON 9.6

NAME ______________________ DATE __________

Practice C

For use with pages 568–574

Solve the equation by using the LCD. Check each solution.

1. $\dfrac{2}{x-10} - \dfrac{3}{x-2} = \dfrac{6}{x^2-12x+20}$

2. $\dfrac{2}{x^2-6x+8} = \dfrac{1}{x-4} + \dfrac{2}{x-2}$

3. $\dfrac{100-4x}{3} = \dfrac{5x+6}{4} + 6$

4. $\dfrac{4}{x-2} - \dfrac{3}{x+1} = \dfrac{8}{x^2-x-2}$

5. $\dfrac{3}{x-8} - \dfrac{4}{x-2} = \dfrac{28}{x^2-10x+16}$

6. $\dfrac{2x}{x-2} - \dfrac{4x-1}{3x+2} = \dfrac{17x+4}{3x^2-4x-4}$

Solve the equation by cross multiplying. Check each solution.

7. $\dfrac{3x-1}{6x-2} = \dfrac{2x+5}{4x-13}$

8. $\dfrac{5x+2}{10x-3} = \dfrac{x-8}{2x+3}$

9. $\dfrac{x}{2x-1} = \dfrac{3}{x+2}$

10. $\dfrac{2}{2x+3} = \dfrac{2}{x-5}$

11. $\dfrac{x}{3x-5} = \dfrac{2}{x-1}$

12. $\dfrac{x+1}{2x-3} = 2$

Solve the equation using any method. Check each solution.

13. $\dfrac{x-2}{3x+5} = 4$

14. $\dfrac{4}{x+2} + \dfrac{1}{x-2} = \dfrac{5x-6}{x^2-4}$

15. $\dfrac{2x}{x+3} + \dfrac{5}{x} - 4 = \dfrac{18}{x^2+3x}$

16. $\dfrac{\left(\dfrac{1}{x} - \dfrac{1}{x+1}\right)}{\left(\dfrac{1}{x+1}\right)} = 2$

17. $\dfrac{\left(\dfrac{4}{x-3} + 3\right)}{\left(\dfrac{4x-1}{x-3} + 4\right)} = 1$

18. $\dfrac{\left(\dfrac{7}{x+1} - \dfrac{3}{x-1}\right)}{\left(\dfrac{2}{x^2-1}\right)} = 3$

19. ***Temperature*** The average monthly high temperature in Jackson, Mississippi can be modeled by $T = -\dfrac{191(t-30)}{t^2-16.5t+114}$ where T is measured in degrees Fahrenheit and $t = 1, 2, \ldots 12$ represents the months of the year. During which month is the average monthly high temperature equal to 57.3° F?

20. ***Average Cost*** You invest \$40,000 to start a nacho stand in a shopping mall. You can make each basket of nachos for \$0.70. How many baskets must you sell before your average cost per basket is \$1.50.

NAME ______________________ DATE ____________

Reteaching with Practice

For use with pages 568–574

GOAL Solve rational equations

VOCABULARY

To solve a rational equation, multiply each term on both sides of the equation by the LCD of the terms. Simplify and solve the resulting polynomial equation.

To solve a rational equation for which each side of the equation is a single rational expression, use **cross multiplying.**

EXAMPLE 1 An Equation with One Solution

Solve: $\frac{7}{x} - \frac{1}{3x} = \frac{5}{3}$.

SOLUTION

The least common denominator is $3x$.

$\frac{7}{x} - \frac{1}{3x} = \frac{5}{3}$	Write original equation.
$3x\left(\frac{7}{x} - \frac{1}{3x}\right) = 3x\left(\frac{5}{3}\right)$	Multiply each side by the LCD, $3x$.
$21 - 1 = 5x$	Simplify.
$20 = 5x$	Subtract.
$4 = x$	Divide each side by 5.

The solution is 4. Check this in the original equation.

Exercises for Example 1

Solve the equation by using the LCD. Check each solution.

1. $\frac{3}{x} - \frac{2}{x+1} = \frac{4}{x}$

2. $\frac{2x}{x+3} - 5 = \frac{1}{x+3}$

3. $\frac{4}{x} - \frac{1}{x+2} = \frac{2}{x}$

NAME ______________________________ DATE __________

Reteaching with Practice

For use with pages 568–574

EXAMPLE 2 An Equation with Two Solutions

Solve: $\dfrac{5x}{x-1} - 2 = \dfrac{14}{x^2-1}$.

SOLUTION

Begin by writing each denominator in factored form.
The LCD is $(x+1)(x-1)$.

$$\frac{5x}{x-1} - 2 = \frac{14}{(x+1)(x-1)}$$

$$(x+1)(x-1)\cdot\frac{5x}{x-1} - (x+1)(x-1)\cdot 2$$

$$= (x+1)(x-1)\cdot\frac{14}{(x+1)(x-1)}$$ Multiply each term by LCD.

$$5x(x+1) - 2(x+1)(x-1) = 14$$ Simplify.

$$5x^2 + 5x - 2x^2 + 2 = 14$$ Distribute and use FOIL.

$$3x^2 + 5x + 2 = 14$$ Combine like terms.

$$3x^2 + 5x - 12 = 0$$ Write in standard form.

$$(3x-4)(x+3) = 0$$ Factor.

$$3x - 4 = 0 \text{ or } x + 3 = 0$$ Use zero product property.

$$x = \frac{4}{3} \qquad x = -3$$

The solutions are $\frac{4}{3}$ and -3. Check these in the original equation.

Exercises for Example 2

Solve the equation by using the LCD. Check each solution.

4. $\dfrac{5x}{x-1} - 3 = \dfrac{2x+5}{x^2-1}$

5. $\dfrac{2x}{x-2} - \dfrac{4x-1}{3x+2} = \dfrac{17x+4}{3x^2-4x-4}$

6. $x - \dfrac{24}{x} = 5$

LESSON
9.6

NAME ____ DATE __________

Quick Catch-Up for Absent Students

For use with pages 568–574

The items checked below were covered in class on (date missed) __________

Lesson 9.6: Solving Rational Equations

____ **Goal 1:** Solve rational equations. (pp. 568–569)

Material Covered:

____ Example 1: An Equation with One Solution

____ Example 2: An Equation with an Extraneous Solution

____ Example 3: An Equation with Two Solutions

____ Example 4: Solving an Equation by Cross-Multiplying

Vocabulary:

cross multiplying, p. 569

____ **Goal 2:** Use rational equations to solve real-life problems. (p. 570)

Material Covered:

____ Example 5: Writing and Using a Rational Model

____ Student Help: Study Tip

____ Example 6: Using a Rational Model

____ Other (specify) ____

____

Homework and Additional Learning Support

____ Textbook (specify) pp. 571–574

____

____ Internet: Extra Examples at www.mcdougallittell.com

____ *Reteaching with Practice* worksheet (specify exercises)__________________

____ *Personal Student Tutor* for Lesson 9.6

Lesson 9.6

NAME ______________________________ DATE __________

Interdisciplinary Application

For use with pages 568–574

Mumps

BIOLOGY Mumps is a disease caused by a virus found in the saliva. Mumps affects the parotid glands or the salivary glands, so this disease is also known as parotitis. These glands, found in the cheeks, become swollen and cause pain in the area between the cheek and the ear.

A person may have mumps and show no symptoms. In that case, a person could pass the disease on to others without knowing. Most cases of the mumps are not life threatening. Mumps may affect other parts of the body, such as the central nervous system, causing serious headaches and tremendously high fever.

As of today there is no medication or drug that affects mumps once the virus has settled in the body. In the 1960s a mumps vaccine was made available which helps prevent the disease from occurring.

In Exercises 1–4, use the following information.

From 1985 through 1996, the number of reported cases of mumps C can be modeled by

$$C = \frac{-136t + 2952}{0.035t^2 - 0.29t + 1}$$

where t is the year, with $t = 0$ corresponding to 1985.

1. In which year(s) is the number of cases of mumps equal to 2952?
2. In which year(s) is the number of cases of mumps less than 1000?
3. Sketch the graph for $0 \le t \le 11$.
4. Using the graph, estimate the year with the most cases.

NAME ______________________________ DATE __________

Math and History Application

For use with page 574

HISTORY Human beings are used to living with the pressure of the atmosphere at sea level, which is about 14 pounds per square inch. In fact, we don't even feel it. But when we explore the depths of the ocean or the heights of the Himalayas, our sensitivity to high and low pressures becomes apparent. Divers experience the direct effect of pressure on their bodies and below a certain depth need to be protected, as William Beebe was in his bathysphere, or Jacques Piccard and Don Walsh were when they descended in their bathyscaphe Trieste to a depth of 35,810 feet in the Marianas Trench in the Pacific. It took Piccard and Walsh about five hours to make the trip to the bottom. In Exercise 4 below you will calculate the pressure that their vessel endured at this depth.

As you increase in elevation the air thins out. Most mountain climbers need oxygen to cope with the effects of low pressure near the summit. See Exercise 5 for a formula that relates pressure to height as you climb above sea level.

MATH In the following problems you will look at how the recommended oxygen percentage and the water pressure depend on a diver's depth, and how atmospheric pressure changes as you climb up from sea level.

1. What is the recommended percent of oxygen in the air that Sylvia Earle was breathing at her record depth of 1250 feet?

2. What does the formula $p = \dfrac{660}{d + 33}$ give for the "recommended" percentage of oxygen in the air you breathe at sea level?

3. As you descend below the ocean surface, the pressure increases by an amount equal to the weight of the column of water above you. This pressure is often measured in atmospheres, where one atmosphere is the pressure at sea level. An approximate formula for pressure P in atmospheres at a depth of d meters is $P = 1 + 0.1d$. Use this formula to find the pressure on Sylvia Earle during her walk at 1250 feet. (1 meter $\approx$ 3.28 feet)

4. The deepest point in the ocean is the Marianas Trench, 11,030 meters below the surface (Piccard and Walsh went to nearly this depth). What is the pressure there in atmospheres?

5. The ratio of the pressure h meters above sea level to the pressure at sea level is given approximately by this formula:

$$\exp\left(-\frac{0.12h}{1000}\right)$$

where exp is the exponential function. How does the pressure on the summit of Mt. Everest (8843 meters) compare with the pressure at sea level?

NAME ______________________ DATE __________

Challenge: Skills and Applications

For use with pages 568–574

In Exercises 1–6, solve each equation.

1. $\frac{x+1}{x^2-4} + \frac{x-1}{x^2+x-2} = \frac{1}{x+2}$

2. $\frac{x+4}{x+3} + \frac{x^2-3}{x^2+2x-3} = \frac{x-3}{x-1}$

3. $\frac{2x^2-1}{x^3+1} + \frac{x+2}{x^2-x+1} = \frac{7}{x+1}$

4. $\frac{4x+1}{(x-2)^2} - \frac{5}{x-2} = \frac{3x+2}{(x-2)^3}$

5. $\frac{4}{x-3} - \frac{x+2}{x^2+3x+9} = \frac{2x^2+6}{x^3-27}$

6. $\frac{2x-5}{x-3} - 3 = \frac{3}{2x^2-7x+3}$

7. Show that the following equation has no solution.

$$\frac{6}{x-1} - \frac{x-x^2}{x^2-4x+3} = \frac{x}{x-3}$$

8. Susan Chen plans to run a 12.2 mile course in 2 hours. For the first 8.4 miles she plans to run at a slower pace, then she plans to speed up by 2 mi/h for the rest of the course. What is the slower pace that Susan will need to maintain in order to achieve this goal?

9. Jorge Martinez is making a business trip by car. After driving half the total distance, he finds he has averaged only 20 mi/h, because of numerous traffic tie-ups. What must be his average speed for the second half of the trip if he is to average 40 mi/h for the entire trip? Answer this question using the following method.

   a. Let d represent the distance Jorge has traveled so far, and let r represent his average speed for the remainder of the trip. Write a rational function, in terms of d and r, that gives the total time Jorge's trip will take.

   b. Write a rational expression, in terms of d and r, that gives his average speed for the entire trip.

   c. Using the expression you wrote in part (b), write an equation expressing the fact that his average speed for the entire trip is 40 mi/h. Solve this equation for r if you can. If you cannot, explain why not.

LESSON 9.6

NAME ________________________ DATE ____________

Quiz 2

For use after Lessons 9.4–9.6

Perform the indicated operation and simplify. *(Lessons 9.4 and 9.5)*

1. $\dfrac{4x^3y^5}{3x^2y^4} \cdot \dfrac{9x^3y^2}{12xy}$

2. $\dfrac{x^2 + 5x + 6}{3x} \div (x + 3)$

3. $\dfrac{5x}{x^2 - x - 12} + \dfrac{3x}{x - 4}$

4. $\dfrac{7x^2}{9x^2 - 25} - \dfrac{2}{6x + 10}$

Simplify the complex fraction. *(Lesson 9.5)*

5. $\dfrac{\dfrac{7}{x} + 11}{\dfrac{2}{3x} - 2}$

6. $\dfrac{25 - \dfrac{1}{x^2}}{\dfrac{1}{5x^2 - x}}$

7. $\dfrac{\dfrac{3}{x^2 - 9} - \dfrac{2}{x + 3}}{\dfrac{1}{6x - 18}}$

8. $\dfrac{\dfrac{3x}{x - 3} + \dfrac{6}{x + 2}}{\dfrac{3}{x^2 - x - 6}}$

Solve the equation and check the solution. *(Lesson 9.6)*

9. $\dfrac{3}{x + 1} = \dfrac{3}{x^2 + x}$

10. $\dfrac{3}{2x} - \dfrac{1}{x - 5} = 1$

Answers

1. ____________
2. ____________
3. ____________
4. ____________
5. ____________
6. ____________
7. ____________
8. ____________
9. ____________
10. ____________

Lesson 9.6

NAME ______________________ DATE __________

Chapter Review Games and Activities

For use after Chapter 9

Solve the problems below. Match the answers with the column on the right. Place the letters that are in front of the answers in the blanks below. When they are on the lines below matched with the problem number, you will be able to answer the following riddle: *How do you measure the speed of a potato?*

1. x and y vary inversely. What is the constant relating x and y when $x = 18$ and $y = \frac{2}{3}$?
2. Using the constant from problem #1, find y when $x = 2$.
3. z varies jointly with x and y. Use the values of $x = 2, y = 7$, and $z = 42$. What is the constant that relates the variables?
4. Using the information from problem #3, find z when $x = 5$ and $y = \frac{3}{5}$.
5. Simplify $\dfrac{x^2 - 4}{2x^2 + 5x - 3} \cdot \dfrac{x + 3}{x - 2}$.
6. Simplify $\dfrac{3x^2 + 5x - 2}{3x^2 + 8x - 3} \cdot \dfrac{4x^2 - 11x + 6}{4x^2 + 5x - 6}$.
7. Simplify $\dfrac{\frac{2}{x} + 6}{\frac{3x + 1}{3x^2}}$.
8. Solve $\dfrac{5}{x + 8} = \dfrac{6}{x - 3}$.
9. Solve $1 + \dfrac{5}{x - 2} = \dfrac{2}{x^2 - 4}$.

(A) $3x - 1$

(B) $k = \dfrac{27}{2}$

(C) $k = 27$

(D) $k = 12$

(E) $\dfrac{x - 2}{x + 3}$

(M) $y = 6$

(N) $3x + 1$

(O) $k = 3$

(P) $z = 9$

(R) $6x$

(S) $x = -63$

(T) $x = 4, -1$

(U) $\dfrac{x + 2}{2x - 1}$

(V) $x + 3$

(W) $3x + 1$

(Y) $2x$

With a ____ ____ ____ ____ ____ ____ ____ ____ ____ ____

8	4	5	1	3	2	6	9	6	7

CHAPTER 9

NAME ______________________ DATE __________

Chapter Test A

For use after Chapter 9

The variables x and y vary inversely. Use the given values to write an equation relating x and y. Then find y when $x = 2$.

1. $x = 1, y = 2$
2. $x = 4, y = -1$
3. $x = 6, y = 2$

The variable z varies jointly with x and y. Use the given values to write an equation relating x, y, and z. Then find z when $x = 2$ and $y = 4$.

4. $x = 2, y = 4, z = 1$
5. $x = -2, y = 1, z = 2$

Graph the function.

6. $y = \frac{1}{x}$

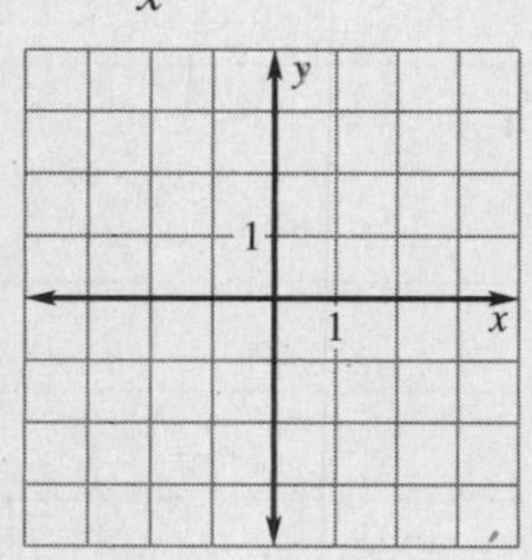

7. $y = \frac{2}{x - 1}$

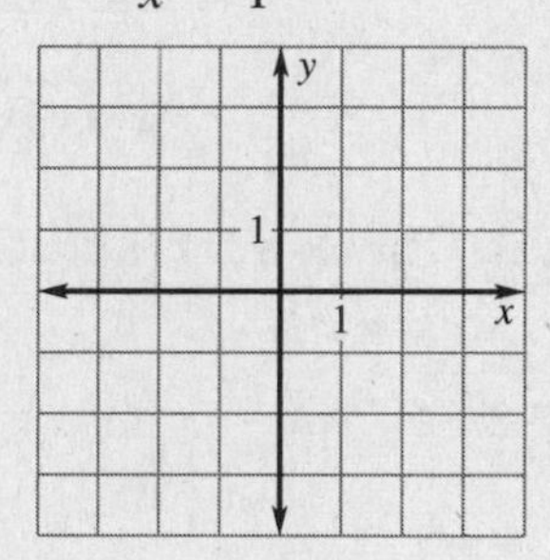

8. $y = \frac{x}{x - 2}$

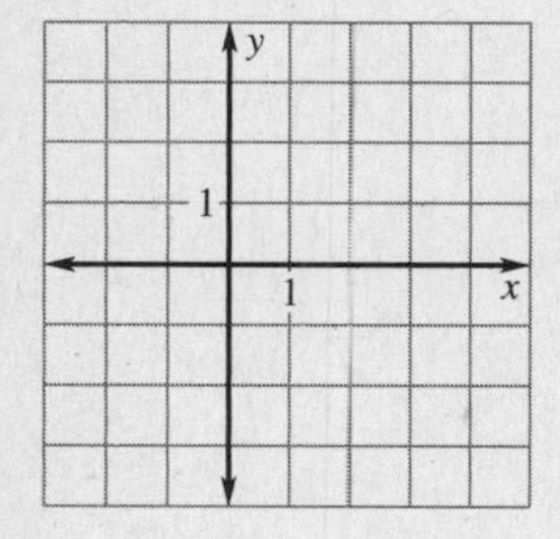

9. $y = \frac{1}{x^2}$

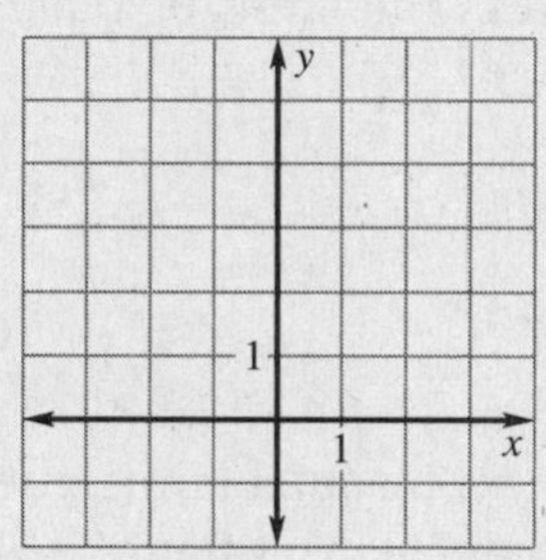

10. $y = x^2$

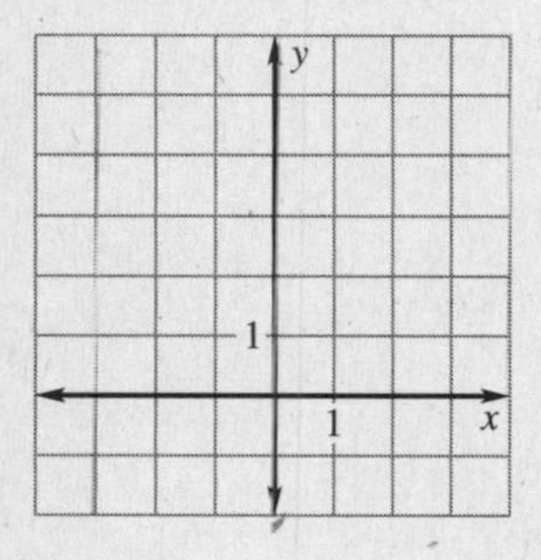

11. $y = \frac{x^2 - 1}{x}$

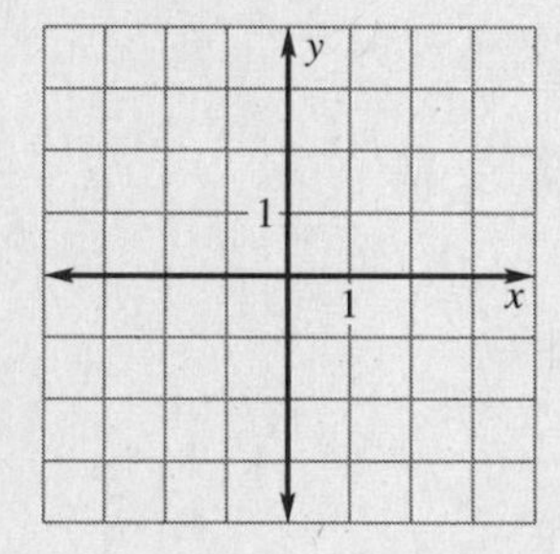

Answers

1. ____________
2. ____________
3. ____________
4. ____________
5. ____________
6. Use grid at left.
7. Use grid at left.
8. Use grid at left.
9. Use grid at left.
10. Use grid at left.
11. Use grid at left.

Review and Assess

NAME ______________________ DATE __________

Chapter Test A

For use after Chapter 9

Perform the indicated operation. Simplify the result.

12. $\frac{x^3}{4} \cdot \frac{2}{x^2}$

13. $\frac{x+1}{x} \cdot \frac{x^3}{(x+1)^2}$

14. $\frac{x+5}{x} \div \frac{x+5}{2x}$

15. $\frac{3x+1}{x-2} + \frac{2x-1}{x-2}$

16. $\frac{5x^2-8x}{x^2-9} - \frac{4x+9x^2}{x^2-9}$

17. $\frac{9x^3}{8x+32} \cdot \frac{2x+8}{-3x^4}$

Simplify the complex fraction.

18. $\frac{5+\frac{1}{4}}{2+\frac{2}{3}}$

19. $\frac{\frac{x}{3}-4}{5+\frac{1}{x}}$

20. $\frac{\frac{x+3}{3x^2}}{\frac{6x^2}{(x+3)^2}}$

Solve the equation using any method. Check each solution.

21. $\frac{3x}{4} = \frac{(x+1)}{2}$

22. $\frac{10}{x+3} + \frac{10}{3} = 6$

23. $\frac{2x-9}{x-7} + \frac{x}{2} = \frac{5}{x-7}$

24. *Geometry Connection* The similar triangles below have congruent angles and proportional sides. Express z in terms of x and y.

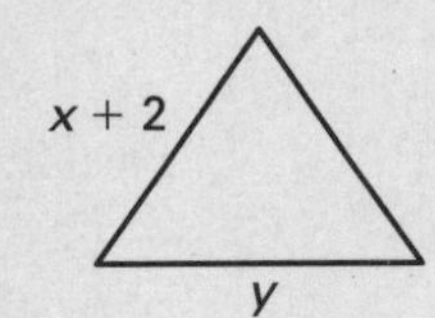

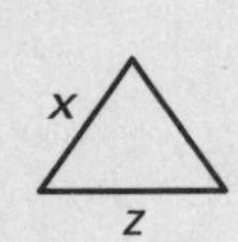

25. *Starting a Business* You start a business manufacturing golf balls, spending \$42,000 for supplies and equipment. You figure it will cost \$12 per dozen to manufacture the golf balls. How many dozens of golf balls must you produce before your average total cost per dozen is \$15?

12. ______
13. ______
14. ______
15. ______
16. ______
17. ______
18. ______
19. ______
20. ______
21. ______
22. ______
23. ______
24. ______
25. ______

CHAPTER 9

NAME ______________________ DATE __________

Chapter Test B

For use after Chapter 9

The variables x and y vary inversely. Use the given values to write an equation relating x and y. Then find y when $x = 4$.

1. $x = -2, y = 2$
2. $x = 8, y = \frac{1}{2}$
3. $x = -\frac{2}{3}, y = 12$

The variable z varies jointly with x and y. Use the given values to write an equation relating x, y, and z. Then find z when $x = -1$ and $y = 4$.

4. $x = 2, y = 4, z = -2$
5. $x = 4, y = \frac{1}{2}, z = -1$

Graph the function.

6. $xy = 1$

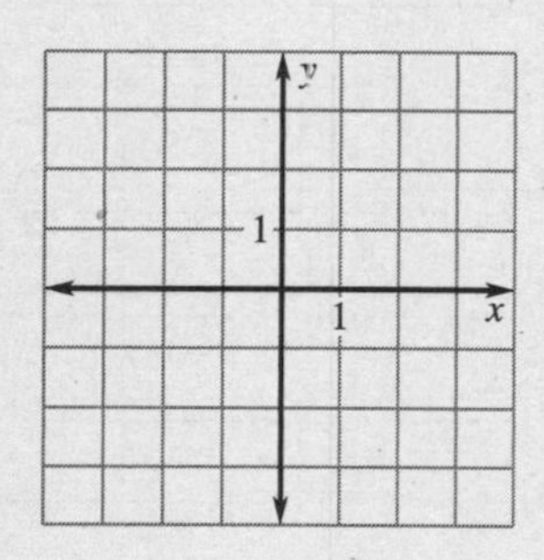

7. $y = \dfrac{3}{x - 1}$

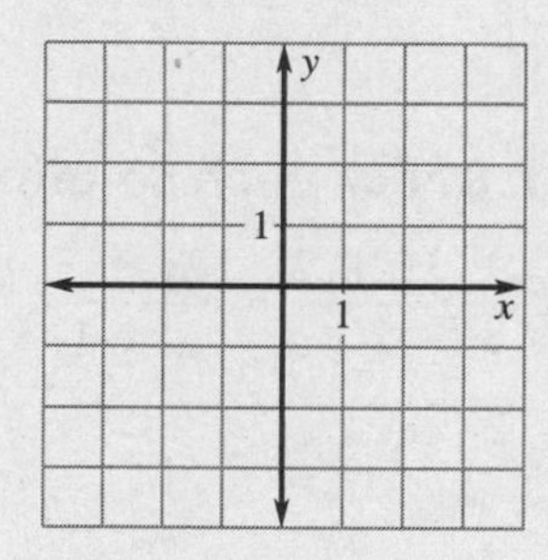

8. $y = \dfrac{x}{x - 4}$

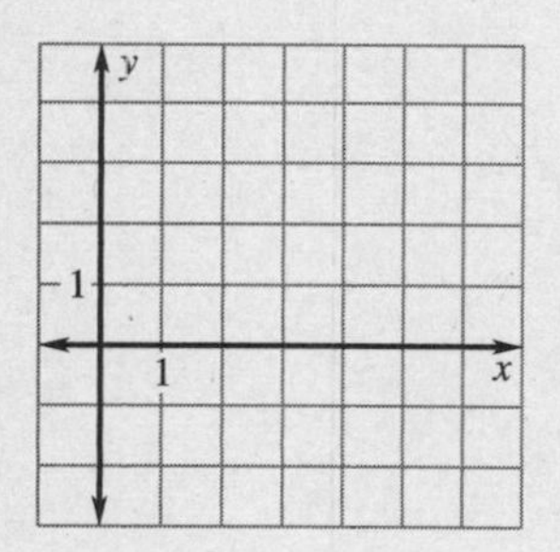

9. $y = \dfrac{1}{2x^2}$

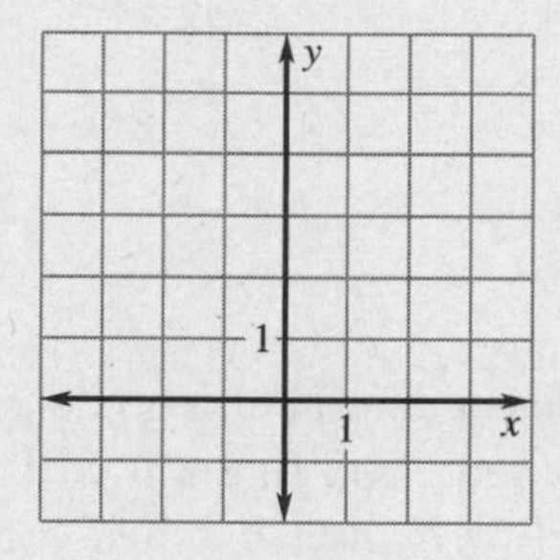

10. $y = x^2 - 1$

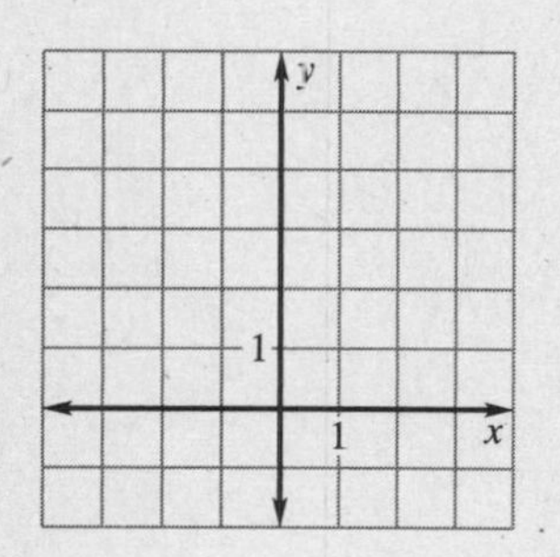

11. $y = \dfrac{-x^2}{x - 1}$

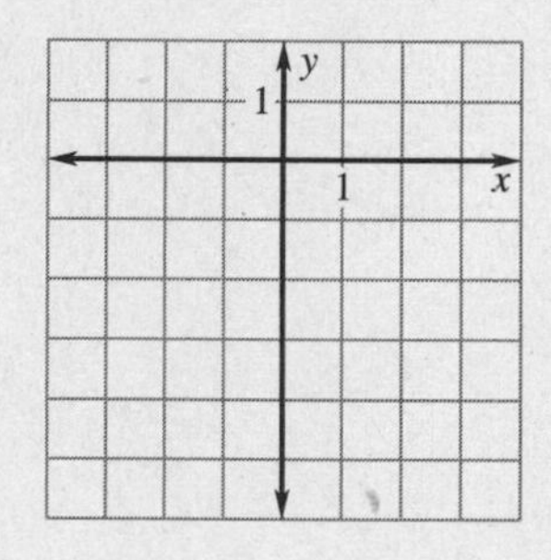

Answers

1. ______________
2. ______________
3. ______________
4. ______________
5. ______________
6. Use grid at left.
7. Use grid at left.
8. Use grid at left.
9. Use grid at left.
10. Use grid at left.
11. Use grid at left.

Review and Assess

NAME ______________________ DATE __________

Chapter Test B

For use after Chapter 9

Perform the indicated operation. Simplify the result.

12. $\dfrac{x}{2y} + \dfrac{5x}{y}$

13. $\dfrac{y}{y-3} - \dfrac{3}{y-3}$

14. $\dfrac{(x+1)^2}{(x-1)} \cdot \dfrac{(2x-2)}{(x+1)}$

15. $\dfrac{(x^2-5x+4)}{x^3} \div \dfrac{(x-4)}{x}$

16. $\dfrac{6x+5}{2x+6} - \dfrac{2x-7}{2x+6}$

17. $\dfrac{x^3-3x^2}{3x+6} \div \dfrac{x^3-8x^2+15x}{6x^2-18x-60}$

Simplify the complex fraction.

18. $\dfrac{\frac{5}{x}}{\frac{x}{5}}$

19. $\dfrac{\frac{x+y}{x-y}}{\frac{x^2+2xy+y^2}{x^2-2xy+y^2}}$

20. $\dfrac{\frac{2}{x}+\frac{3}{xy}}{\frac{3}{x^2}+\frac{1}{5}}$

Solve the equation using any method. Check each solution.

21. $\dfrac{3}{x+1} = \dfrac{2}{x-4}$

22. $\dfrac{x^2+2x+2}{x-1} = \dfrac{2x+3}{x-1}$

23. $\dfrac{7}{4} + \dfrac{2}{3}y^2 = \dfrac{13y}{6}$

24. ***Geometry Connection*** The similar triangles below have congruent angles and proportional sides. Express z in terms of x and y.

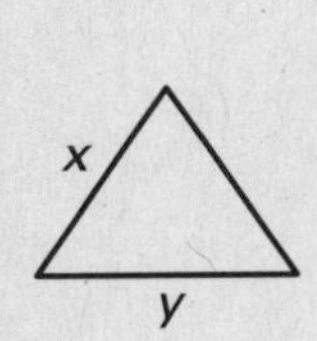

25. ***Starting a Business*** You start a business manufacturing hats, spending $8,000 for supplies and equipment. You figure it will cost $4.95 per hat to manufacture the hats. How many hats must you produce before your average total cost per hat is $5?

12. ______________
13. ______________
14. ______________
15. ______________
16. ______________
17. ______________
18. ______________
19. ______________
20. ______________
21. ______________
22. ______________
23. ______________
24. ______________
25. ______________

CHAPTER 9

NAME ______________________ DATE __________

Chapter Test C

For use after Chapter 9

The variables *x* and *y* vary inversely. Use the given values to write an equation relating *x* and *y*. Then find *y* when *x* = 3.

1. $x = -1, y = 9$
2. $x = \frac{3}{5}, y = \frac{5}{3}$
3. $x = -6, y = \frac{1}{3}$

The variable *z* varies jointly with *x* and *y*. Use the given values to write an equation relating *x*, *y*, and *z*. Then find z when *x* = 2 and *y* = 3.

4. $x = 4, y = 2, z = -1$
5. $x = \frac{1}{2}, y = \frac{1}{3}, z = \frac{3}{2}$

Graph the function.

6. $xy = 2$

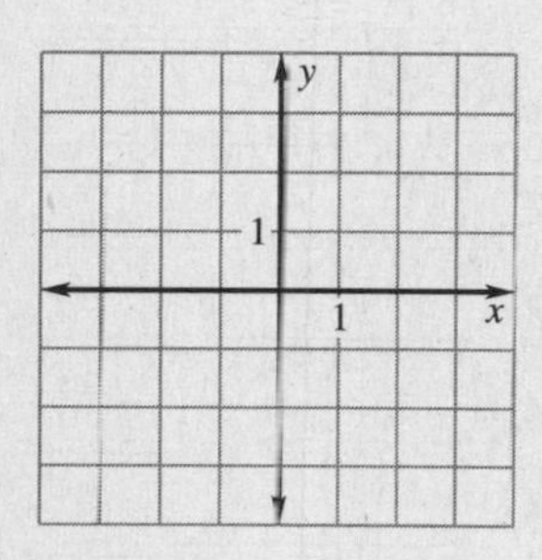

7. $y = \dfrac{2}{x - 2}$

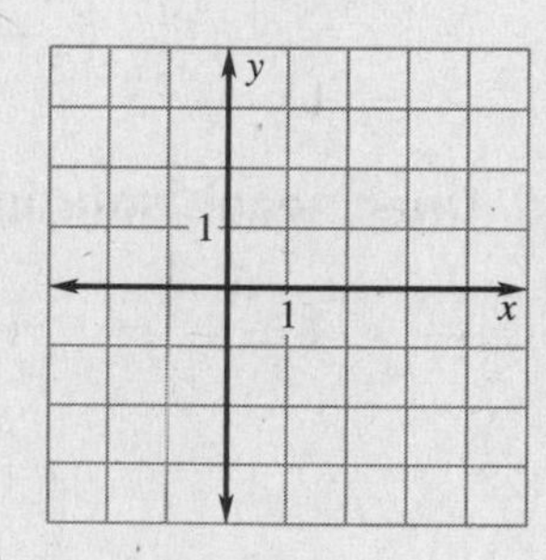

8. $y = \dfrac{2x}{x - 4}$

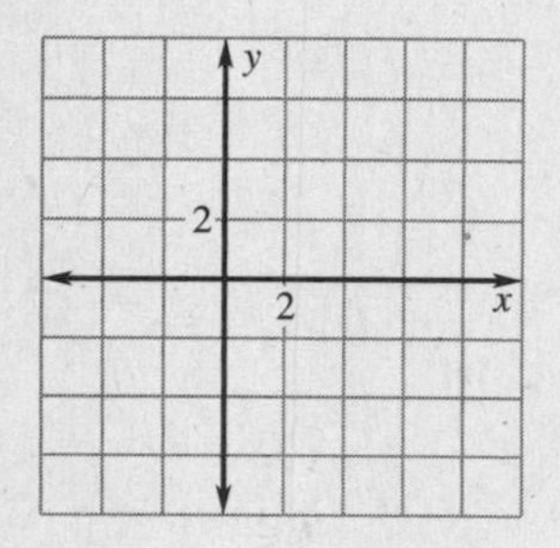

9. $y = \dfrac{4}{x^2}$

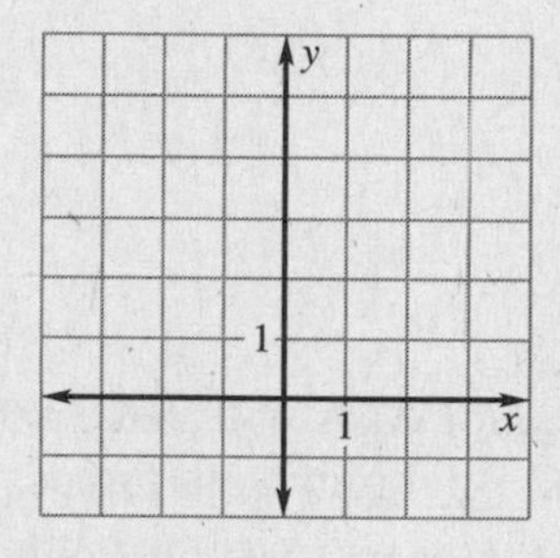

10. $y = \dfrac{-2x^2}{x - 2}$

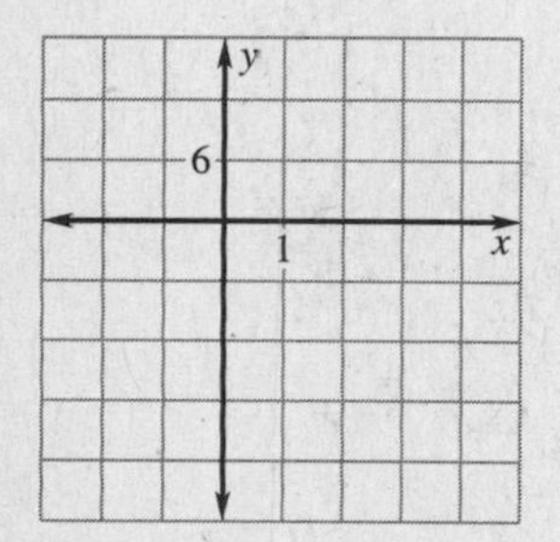

11. $y = \dfrac{x^2 - 3x + 5}{x + 1}$

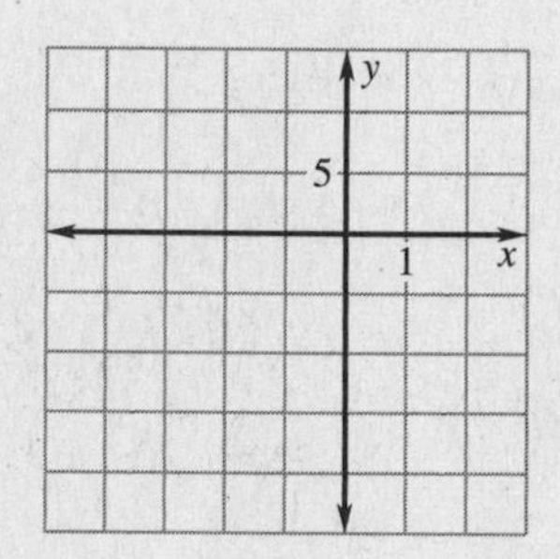

Answers

1. ______________
2. ______________
3. ______________
4. ______________
5. ______________
6. Use grid at left.
7. Use grid at left.
8. Use grid at left.
9. Use grid at left.
10. Use grid at left.
11. Use grid at left.

Review and Assess

CHAPTER 9 CONTINUED

NAME ______________________ DATE __________

Chapter Test C

For use after Chapter 9

Perform the indicated operation. Simplify the result.

12. $\frac{5}{x} + \frac{3}{y} - \frac{2}{z}$

13. $\frac{3x-5}{x^2-9} + \frac{7-x}{9-x^2} - \frac{1}{x+3}$

14. $\frac{3y-5}{2y-6} - \frac{4y-2}{5y-15}$

15. $(30x^3 - 6x^2) \div \frac{15x^2-3x}{4x^2+4x-24}$

16. $\frac{3x}{2x-3} + \frac{3x+6}{2x^2+x-6}$

17. $\frac{6x^2+x-2}{6x^2+7x+2} \cdot \frac{2x^2+9x+4}{4-7x-2x^2}$

Simplify the complex fraction.

18. $\dfrac{\frac{1}{5x^2} - \frac{2}{y}}{\frac{7}{10x} + \frac{3}{2y^2}}$

19. $\dfrac{\left(\frac{10}{x+1}\right)}{\left(\frac{1}{2} + \frac{3}{x+1}\right)}$

20. $2 - \dfrac{2}{2 - \left(\frac{2}{2-x}\right)}$

Solve the equation using any method. Check each solution.

21. $\frac{2}{x} = \frac{x}{x^2-8}$

22. $\frac{3}{x-2} = \frac{4}{x-3} - \frac{6}{x^2-5x+6}$

23. $\frac{3}{x+1} + \frac{x-2}{3} = \frac{13}{3x+3}$

24. ***Geometry Connection*** The similar triangles below have congruent angles and proportional sides. Express z in terms of x and y.

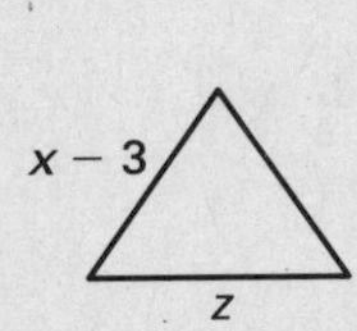

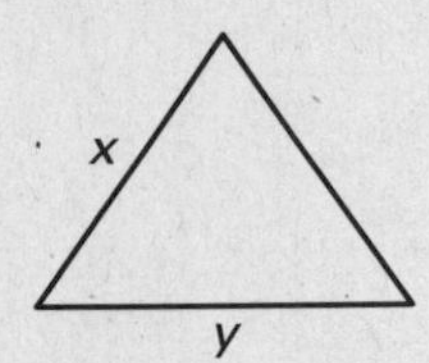

25. ***Fund Raiser*** As a fund raiser, your junior class will make and sell holiday greeting cards. You spend \$750 as an initial startup cost. It will cost you \$4.25 per box to print, and you will sell the cards at \$10 per box. How many boxes must you sell to show a profit?

12. ____________
13. ____________
14. ____________
15. ____________
16. ____________
17. ____________
18. ____________
19. ____________
20. ____________
21. ____________
22. ____________
23. ____________
24. ____________
25. ____________

Review and Assess

CHAPTER 9

NAME ______________________ DATE ________

SAT/ACT Chapter Test

For use after Chapter 9

1. The variable x varies inversely with y. When $x = -3$, $y = -2$. Which equation relates x and y?

Ⓐ $\frac{x}{y} = \frac{-3}{-2}$ Ⓑ $xy = 6$

Ⓒ $\frac{x}{y} = \frac{3}{2}$ Ⓓ $x = 6y$

2. The variable z varies jointly with x and y. When $x = 5$ and $y = 2$, $z = 10$. Which equation relates x, y, and z?

Ⓐ $z = xy$ Ⓑ $z = \frac{1}{10}xy$

Ⓒ $z = \frac{x}{y}$ Ⓓ $z = 10xy$

3. Which function is graphed?

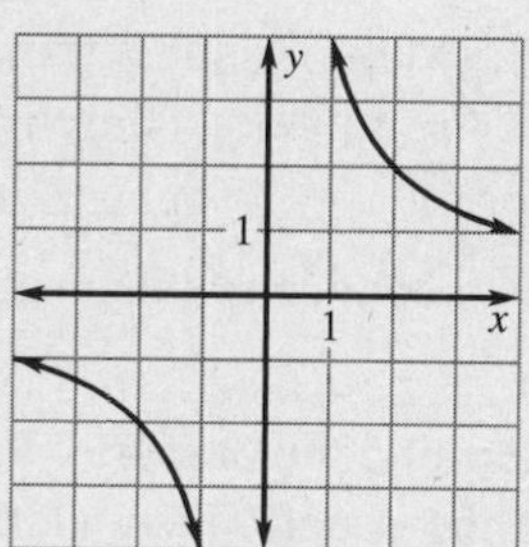

Ⓐ $4xy = 0$ Ⓑ $y = \frac{x}{4}$

Ⓒ $x = \frac{y}{4}$ Ⓓ $xy = 4$

4. What are the solutions of the equation $\frac{x}{1} = \frac{2}{x - 1}$?

Ⓐ $-2, 1$ Ⓑ $2, 1$

Ⓒ $-1, 2$ Ⓓ $-1, -2$

5. What is the sum $\frac{x + 1}{x} + \frac{x}{2}$?

Ⓐ $\frac{x(x + 1)}{2x}$ Ⓑ $\frac{x^2 + 2x + 2}{x + 2}$

Ⓒ $\frac{x^2 + x + 2}{x + 2}$ Ⓓ $\frac{x^2 + 2x + 2}{2x}$

6. What is the simplified form of the complex fraction $\frac{\frac{1}{x}}{\frac{x}{x^2 + 1}}$?

Ⓐ $\frac{x^2 + 1}{x^2}$ Ⓑ $\frac{1}{x^2}$

Ⓒ 2 Ⓓ $\frac{x^2 + 1}{2x}$

7. What is the quotient $\frac{(x^2 + 5x + 4)}{(x + 1)} \div \frac{(x + 4)}{(x + 1)}$?

Ⓐ $\frac{(x + 4)^2(x + 1)}{(x + 1)^2}$ Ⓑ $\frac{x + 1}{x + 4}$

Ⓒ $\frac{(x + 4)^3}{(x + 1)^2}$ Ⓓ $x + 1$

8. What is the product $\frac{(x + 1)^2}{x^3} \cdot \frac{x^2 + x}{x + 1}$?

Ⓐ $\frac{x^2 + 1}{x^2 + x + 1}$ Ⓑ $\frac{(x + 1)^3}{(x + 1)}$

Ⓒ $\frac{x^2 + 2x + 1}{x^2}$ Ⓓ $\frac{x^2 + x + 1}{x^2}$

Quantitative Comparision Choose the statement that is true about the given quantities.

Ⓐ The quantity in column A is greater.
Ⓑ The quantity in column B is greater.
Ⓒ The two quantities are equal.
Ⓓ The relationship cannot be determined from the given information.

9.

Column A	*Column B*
The solution of $\frac{x + 1}{x} = \frac{1}{2}$	The solution of $\frac{x + 1}{x} = \frac{2}{1}$

Ⓐ Ⓑ Ⓒ Ⓓ

Review and Assess

NAME ______________________ DATE ________

Alternative Assessment and Math Journal

For use after Chapter 9

JOURNAL 1. Performing operations on rational functions is an important part of Chapter 9. It is often helpful to review the strategies for simplifying different types of rational expressions such as the following.

i. $\dfrac{A+B}{C}$ *ii.* $\dfrac{C}{C+D}$ *iii.* $\dfrac{A}{C(C+D)}$ *iv.* $\dfrac{C+\dfrac{A}{B}}{B+\dfrac{A}{C}}$

Use the above rational expressions to answer the questions. Present a general process rather than specific answers to solve the particular problem. (a) Write a paragraph to explain how to add *i* and *ii*. (b) How does adding *i* and *iii* differ from part a? (c) Write a paragraph to explain how to multiply *i* and *ii*. (d) Explain how division varies from multiplication with rational expressions.

MULTI-STEP PROBLEM 2. Use the following equations to answer the questions below.

I. $\dfrac{x}{x+3} - \dfrac{2}{x+3} = 4$ **II.** $\dfrac{2x+8}{x+3} - \dfrac{2}{x+3} = x$

a. Solve each of the equations. In which equation did an extraneous root occur? What part in the solution of the equation caused this to happen?

b. In general, in what cases do extraneous roots occur when solving rational equations? Why?

c. Sketch the graph of each function. Determine the domain. Describe what happens at points where the equation is undefined.

d. Use a table of values to examine the functions. At what points in equations I and II is the function undefined?

e. Create and solve a rational function that does not have extraneous roots. Create and solve a rational function with extraneous roots in its solution. State very clearly why you could expect extraneous roots in your solution.

3. ***Critical Thinking*** Give an example of another type of equation in which extraneous roots appear. How is the solution process to solving this equation similar to solving rational equations? How does graphing assist in determining the roots?

Alternative Assessment Rubric

For use after Chapter 9

JOURNAL SOLUTION

1. a–d. Complete answers should address these points:
   - a. • Explain that a common denominator must be found by multiplying the denominators.
   - b. • Explain that a common denominator in this problem would be the same as part a, except that only i would need to change.
   - c. • Explain that in multiplication the numerators and denominators are multiplied and the result is simplified by canceling/reducing common factors.
   - d. • Explain that when dividing, you must multiply the first fraction by the reciprocal of the second fraction.

MULTI-STEP PROBLEM SOLUTION

2. a. Equation I solution is $x = -\frac{14}{3}$; equation II solution is $x = 2$; an extraneous root occurred in equation II; multiplying through by $(x + 3)$ caused this extraneous root to occur.
   - b. *Sample answer:* Extraneous roots can occur when multiplying both sides of an equation by an expression. The number that would make the expression zero cannot be multiplied through, making it an "illegal" operation for that value. This value can become an extraneous root when it is not part of the domain of the original function.
   - c. Domain of I and II: all real numbers except $x = -3$, when the function is undefined, vertical asymptotes occur on the graph.
   - d. In I and II the function is undefined when $x = -3$.
   - e. Answers will vary.

3. *Sample answer:* In solving radical equations, when both sides of the equation are squared, an extraneous root can occur. This is similar to rational equations when there are variables on both sides of the equation. The sides are multiplied by themselves, creating a multi-solution equation. Graphing can help to show where the two graphs intersect. It also illustrates any undefined values.

MULTI-STEP PROBLEM RUBRIC

4 Students complete all parts of the questions accurately. Explanations are clear and logical, showing an understanding of extraneous roots. Student examples illustrate equations both with and without extraneous roots, with clear explanations of why.

3 Students complete the questions and explanations. Explanations may be somewhat vague, but show some understanding of extraneous roots. Student examples illustrate equations both with and without extraneous roots.

2 Students complete questions and explanations. Explanations do not fit the questions. Examples are incomplete or inaccurate.

1 Students' work is very incomplete. Solutions or reasoning are incorrect. Examples are incomplete or inaccurate.

Review and Assess

CHAPTER 9

NAME ______________________ DATE ____________

Project: The Ideal Container

For use with Chapter 9

OBJECTIVE **Determine the best shape and dimensions of a container that will hold a specified volume of liquid.**

MATERIALS graphing calculator or graphing software

INVESTIGATION Have you ever wondered why containers in a grocery store have different shapes and are made of a variety of materials? The people who design food containers must consider many different criteria. They must look at serving size, cost of materials, and shape as it relates to shipping in larger boxes.

Suppose you are part of a package design team. Your job is to design a container that will hold 500 cm^3 of tomato juice.

1. The most desirable package design will minimize material costs. One way to reduce the cost of the material used to make a container is to select the shape with the least surface area for a specified volume. Determine the radius and height of a 500-cm^3 cylindrical can with the least surface area. What is the surface area?
2. Juice is frequently packaged in a rectangular prism-shaped carton. If the base of the prism is a square, what are the dimensions and surface area of the rectangular prism with the least surface area?
3. How would your answer to the previous question change if you assume that the base is a rectangle whose length is twice its width?
4. Investigate other shapes for your tomato juice container. Analyze surface areas for unusual container shapes, such as a cone, pyramid, triangular prism, or sphere.
5. Was the cylinder the most efficient design possible? Why do you think food manufacturers use cylindrical cans so often?
6. Now consider the cylinder again. Suppose that the material for the curved surface of the cylindrical can costs twice as much per cm^2 as the material for the top and bottom. In this situation, what are the radius and height of the least expensive can?
7. Name some other factors that might play a role in determining the best packaging to use for a food product.

PRESENT YOUR RESULTS Present your report as a recommendation to the management of your company. Describe your procedures, and provide data that support the conclusions you make.

NAME ______________________ DATE __________

Project: Teacher's Notes

For use with Chapter 9

GOALS
- Use rational functions and geometry concepts to analyze a real-world problem.

MANAGING THE PROJECT To get students started with this project, you may want to refer them to page 549 of the pupil book before you have started teaching Lesson 9.3. Students will be able to follow the example without having looked at Lessons 9.1 and 9.2.

Encourage students to be creative in thinking about shapes that could be used for containers. Ask them which container they think will have the least surface area.

You may wish to have students extend the project. Here are some ideas:

- Have students consider the problems of packaging objects with unusual shapes. How would you package a football, a boomerang, or a dozen coat hangers?
- Most products are shipped in large quantities after packaging. What issues need to be considered when making larger packages from smaller ones? For example, how would you package 24 cans of tomato juice so that they can be shipped?

RUBRIC **The following rubric can be used to assess student work.**

4 The student answers all questions completely and accurately, and analyzes a wide variety of unusual container shapes. The student's report to management is neat and contains logical arguments as well as supporting evidence such as accurate graphs.

3 The student answers most questions completely and accurately, and analyzes a variety of unusual container shapes. The student's report to management contains logical arguments and supporting evidence. However, the student may not perform all calculations accurately or may not fully address the other factors that play a role in designing packaging.

2 The student answers the questions and analyzes some unusual container shapes. However, the student's work may be incomplete or reflect misunderstandings. For example, the student may not know how to use the cost information in Question 6. The report may indicate a limited grasp of certain ideas or may lack key supporting evidence.

1 The student's answers to some questions are missing or do not show an understanding of key ideas. The report doesn't give a reasonable decision or fails to support the decision.

CHAPTER 9

NAME ______________________ DATE __________

Cumulative Review

For use after Chapters 1–9

Solve the equation or inequality. (1.7)

1. $|3x - 1| = 8$
2. $|12 + 2x| = 5$
3. $\left|\frac{1}{4}x - 6\right| = 8$
4. $|4n - 2| > 6$
5. $|6 - 3n| \le 18$
6. $|11 + x| \ge 18$

Graph the equation. (2.3)

7. $y = \frac{3}{4}x + 2$
8. $y = -2x + 6$
9. $-5x + 3y = 15$
10. $x = 0$
11. $y = -5$
12. $3x + 6y = 18$

Match the equation with its graph. (2.8)

13. $y = |x + 3| - 3$
14. $y = |x| + 3$
15. $y = |x + 3|$

A.
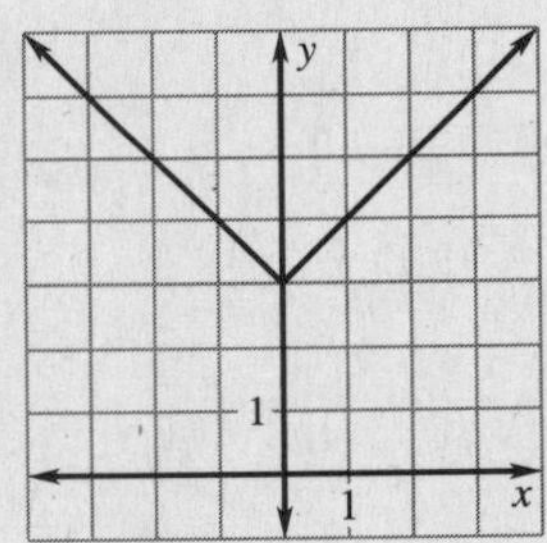

B.

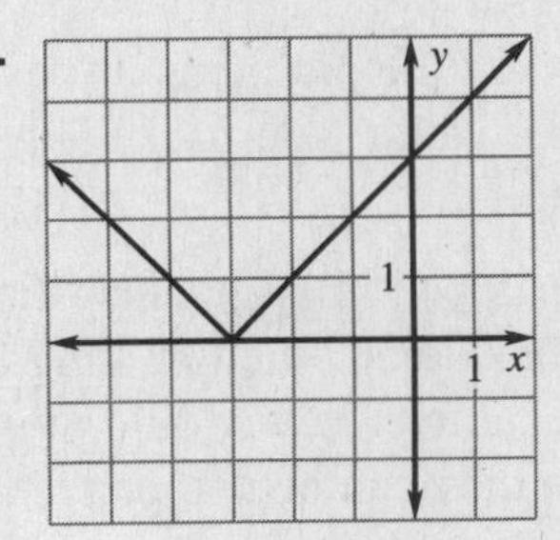

C.
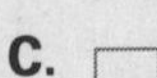
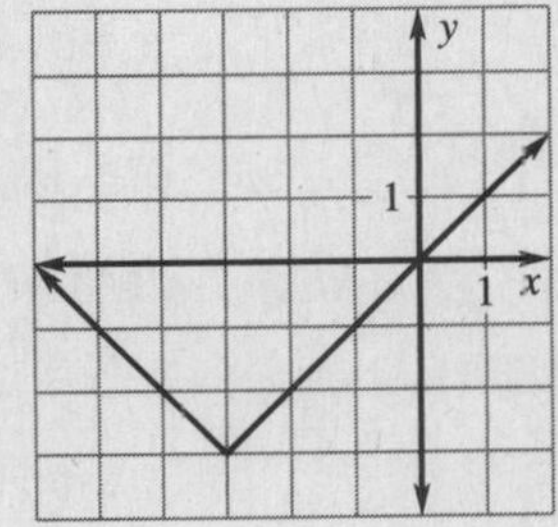

Tell how many solutions the system has. (3.1)

16. $x + 2y = 6$
    $3x - 18 = -6y$
17. $2x - y = 5$
    $2y - 4x = 20$
18. $3x - 4y = 5$
    $x + 2y = 9$

Graph the system of inequalities. (3.3)

19. $2x + y > 6$
    $y < x$
    $x > 4$
20. $x - y \le 0$
    $x + y \ge 8$
    $y \le 6$
21. $5x - 3y \le 15$
    $3x + y \le -3$
    $x \le 0$

Perform the indicated operations. (4.1, 4.2)

22. $\begin{bmatrix} 2 & -3 \\ 0 & 4 \\ -2 & 4 \end{bmatrix} + 4\begin{bmatrix} 3 & -1 \\ 2 & 2 \\ 1 & -5 \end{bmatrix}$
23. $\begin{bmatrix} -3 & 3 \\ 1 & 3 \\ 5 & 0 \end{bmatrix} \cdot \begin{bmatrix} 1 & 3 \\ 4 & -2 \end{bmatrix}$

Factor the expression. (5.2)

24. $x^2 + 7x + 6$
25. $9x^2 - 45x + 50$
26. $9x^2 - 12x + 3$
27. $49x^2 - 81$
28. $9y^2 - 30y + 16$
29. $x^2 + 32x$

Find the absolute value of the complex number. (5.4)

30. $3 + 5i$
31. $-3i$
32. $16 + i$
33. $-2 - i$
34. $-3 - 3i$
35. $-1 + 4i$

Review and Assess

CHAPTER 9 CONTINUED

NAME ______________________ DATE ______

Cumulative Review

For use after Chapters 1–9

Use the quadratic formula to solve the equation. (5.6)

36. $x^2 + 4x = -18$ **37.** $-x^2 + 3x - 4 = 0$ **38.** $-x^2 + 2x = -3$

39. $16x + x^2 = 11x + 3$ **40.** $2(c - 1)^2 - 4 = 1$ **41.** $x^2 - 6x + 9 = 1$

Write a quadratic function in vertex form whose graph has the given vertex and passes through the given point. (5.8)

42. vertex: $(-4, 6)$ point: $(-2, 18)$ **43.** vertex: $(1, -9)$ point: $(-2, -27)$ **44.** vertex: $(0, 3)$ point: $(-5, -22)$

Simplify the expression. (6.1)

45. $\frac{x^5y^3}{x^{-3}y^0}$ **46.** $\frac{2x^2y}{6x^3y^{-4}}$ **47.** $(4x^2y^{-3})^{-2}$

48. $\frac{xy^8}{3x^{-2}} \cdot \frac{6x^3y^4}{2x^3y^{-2}}$ **49.** $-3x^{-2}y^0$ **50.** $\frac{x^4}{x^{-2}}$

Use synthetic substitution to evaluate. (6.2)

51. $f(x) = 2x^3 + x^2 - 2x + 1, x = 2$ **52.** $f(x) = 4x^4 - 3x^2 + 5x - 1, x = -3$

53. $f(x) = x^4 + x^2 - 5x + 11, x = -1$ **54.** $f(x) = -3x^3 + x^2 - x - 3, x = 2$

Decide whether the function is a polynomial function. If it is, write the function in standard form and state the degree, type, and leading coefficient. (6.2)

55. $f(x) = 13 - 6x$ **56.** $f(x) = 2x + \frac{1}{3}x^3 + 8$ **57.** $f(x) = -6x^2 + x - \frac{2}{x}$

Let $f(x) = x^2 + 6x$ and $g(x) = x^2 - 9$. Perform the indicated operation and state the domain. (7.3)

58. $f(x) + g(x)$ **59.** $f(x) - g(x)$ **60.** $f(x) + f(x)$

61. $f(x) \cdot g(x)$ **62.** $f(g(x))$ **63.** $g(g(x))$

Describe how to obtain the graph of g from the graph of f. (7.5)

64. $g(x) = \sqrt{x + 8}, f(x) = \sqrt{x}$ **65.** $g(x) = \sqrt{-x + 6}, f(x) = \sqrt{x}$

66. $g(x) = \sqrt{x + 1} + 5, f(x) = \sqrt{x}$ **67.** $g(x) = 5\sqrt{x - 1}, f(x) = 5\sqrt{x}$

Identify the y-intercept and the asymptote of the graph of the function. (8.1)

68. $y = 6^x$ **69.** $y = -3 \cdot 5^x$ **70.** $y = 2^x + 3$

71. $y = 2^{x-1}$ **72.** $y = 3 \cdot 2^{x-1}$ **73.** $y = 5^{x+1} + 5$

Simplify the expression. (8.3)

74. $(3e^{-2x})^{-1}$ **75.** $\sqrt[3]{27e^{9x}}$ **76.** $e^x \cdot 3e^{2x+1}$

Use a calculator to evaluate the expression. Round the result to three decimal places. (8.4)

77. $\ln 11$ **78.** $\log \sqrt{3}$ **79.** $\log 23.724$

Review and Assess

ANSWERS

Chapter Support

Parent Guide

9.1: about 109 lb **9.2:** *Sample answer:* The graph is a hyperbola through the points $\left(0, \frac{1}{9}\right)$, $\left(2, \frac{5}{3}\right)$, $\left(\frac{1}{3}, 0\right)$, and $\left(\frac{8}{3}, 1\right)$, with vertical asymptote $x = \frac{3}{2}$ and horizontal asymptote $y = \frac{1}{2}$.

9.3: x-intercepts: 1, -5; vertical asymptotes: $x = -3$ and $x = -2$; horizontal asymptote: $y = 1$ **9.4:** with the x-by-$3x$ side down

9.5: $\dfrac{-4x + 15}{(x + 3)(x - 2)}$ **9.6:** 9 weeks

Prerequisite Skills Review

1. $y = 2x$ **2.** $y = -\frac{1}{3}x$ **3.** $y = -0.2x$
4. $12x - 18$ **5.** $-x^5 - 14x^3 - 49x$
6. $x^3 + 5x^2 - 24x$ **7.** $(x - 6)(x + 2)$
8. $(3x - 5)(x + 4)$ **9.** $6(x - 2)(x^2 + 2x + 4)$
10. -3 and 10 **11.** -6 and -2
12. -2, 1, and 3

Strategies for Reading Mathematics

1. a. 420 **b.** $72xy$ **c.** $(x + 1)(x - 1)(x - 2)$
2. a. $\dfrac{71}{210}$ **b.** $\dfrac{x^2 + 12}{24xy}$
c. $\dfrac{4x^2 - 7x + 4}{(x + 1)(x - 1)(x - 2)}$

Lesson 9.1

Warm-Up Exercises

1. $y = 2 - x$ **2.** $y = \dfrac{8}{x}$ **3.** $y = \dfrac{x^2}{2}$

4. $y = \dfrac{0.1}{x}$ **5.** $y = \dfrac{x}{8}$

Daily Homework Quiz

1. 5.138

2. asymptotes: $y = 0$, $y = 5$; y-intercept: 2.5; point of maximum growth: (0, 2.5)

3. $-\dfrac{\ln 3.25}{3} \approx -0.39$

Lesson Opener

Allow 10 minutes.

1. It will decrease. **2.** It will increase.
3. increase the number of people
4. It will double. **5.** It will be reduced to one third of the original amount. **6.** It will stay the same. **7.** Reduce the number of weeks by half, or double the number of people.

Graphing Calculator Activity

1. **a.** 40 **b.** 33.3 **c.** 2 **d.** 1

3. **a.** 6 **b.** 3.75 **c.** 10 **d.** 6

Practice A

1. direct variation **2.** inverse variation
3. neither **4.** inverse variation
5. $y = \dfrac{8}{x}$; 2 **6.** $y = -\dfrac{9}{x}$; $-\dfrac{9}{4}$
7. $y = -\dfrac{36}{x}$; -9 **8.** $y = \dfrac{12}{x}$; 3
9. $y = \dfrac{4}{x}$; 1 **10.** $y = \dfrac{5}{x}$; $\dfrac{5}{4}$ **11.** direct variation
12. neither **13.** inverse variation **14.** direct variation **15.** $z = 3xy$; -36 **16.** $z = \frac{2}{3}xy$; -8

Lesson 9.1 *continued*

17. $z = 2xy; -24$ **18.** $z = -\frac{1}{3}xy; 4$
19. $k = 0.055$ **20.** $I = 0.055Pt$ **21.** $220.00

Practice B

1. direct variation **2.** direct variation
3. inverse variation **4.** neither
5. direct variation **6.** neither
7. $y = \frac{54}{x}; 18$ **8.** $y = \frac{4}{x}; \frac{4}{3}$ **9.** $y = \frac{3}{x}; 1$
10. $z = \frac{3}{4}xy; -\frac{9}{2}$ **11.** $z = 32xy; -192$
12. $z = 3xy; -18$ **13.** $k = 0.035$
14. $I = 0.035Pt$ **15.** $612.50 **16.** $k = 12.84$
17. $PV = 12.84$ **18.** 10.7 cubic liters

Practice C

1. direct variation **2.** inverse variation
3. direct variation **4.** direct variation
5. inverse variation **6.** neither
7. $y = \frac{18}{x}; 3$ **8.** $y = \frac{0.3}{x}; \frac{1}{20}$ **9.** $y = \frac{1}{5x}; \frac{1}{30}$
10. $z = 12xy; -144$ **11.** $z = 16xy; -192$
12. $z = \frac{11}{6}xy; -22$ **13.** $k = 150{,}000$
14. $dp = 150{,}000$ **15.** 12,500 units
16. $k = 0.22$ **17.** $H = 0.22mT$
18. 3.9424 kilocalories

Reteaching with Practice

1. inverse variation **2.** neither
3. direct variation **4.** direct variation
5. $y = \frac{20}{x}; 5$ **6.** $y = \frac{-9}{x}; -2.25$
7. $y = \frac{16}{x}; 4$ **8.** $z = 2xy; 32$
9. $z = -\frac{1}{3}xy; -\frac{16}{3}$ **10.** $z = 32xy; 512$

Real-Life Application

1.

Insulation level	R-10	R-20	R-28	R-40	R-60
RSI value	1.8	3.5	4.9	7.0	10.6
Heat loss	933.3	480.0	342.9	240.0	158.5

2. inverse variation

3.

Outside temperature	10° C	5° C	0° C	−5° C	−10° C
Heat loss	136.7	205.1	273.5	341.8	410.2

4. 0.5 **5.** Decrease the amount of heat loss.

Challenge: Skills and Applications

1. a. P varies directly with T and inversely with V. **b.** V varies inversely with P and directly with T. **c.** T varies jointly with P and V

2. x varies inversely with z: $x = ky$; $y = \frac{m}{z}$; $x = k\left(\frac{m}{z}\right) = \frac{km}{z}$.

3. x varies directly with z: $x = \frac{k}{y}$; $y = \frac{m}{z}$; $x = \frac{k}{m/z} = \frac{k}{m}z$.

4. x varies directly with z: $x = ky^2w$; $y = mz$; $w = \frac{n}{z}$; $x = k(mz)^2 \cdot \frac{n}{z} = km^2nz$.

5. a. The weight varies directly with the radius.
b. 95.6 lb

6. a.

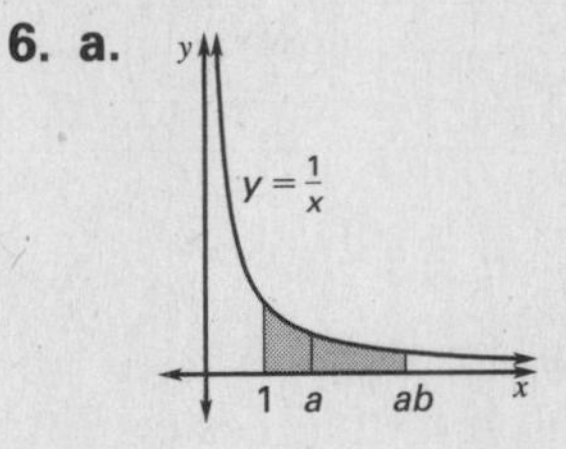

b. $A_1^{ab} = A_1^a + A_a^{ab} = A_1^a + A_1^b$, because $\frac{ab}{a}$ represents the ratio of the two endpoints of A_a^{ab} and it equals $\frac{b}{1}$. So $A_a^{ab} = A_1^b$.

Lesson 9.2

Lesson 9.2

Warm-Up Exercises

1. shifted 1 unit to the right **2.** shifted one unit up **3.** reflected in the x-axis **4.** shifted 2 units to the left and one unit up **5.** y-values are doubled for each corresponding x-value.

Daily Homework Quiz

1. inverse **2.** direct **3.** $y = \dfrac{24}{x}$ **4.** $z = \dfrac{xy}{5}$

5. $z = \dfrac{ky}{x^2}$

Lesson Opener

Allow 10 minutes.

1.

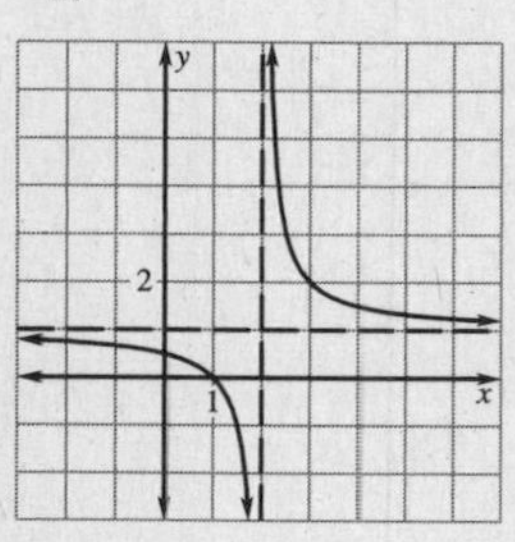

2.

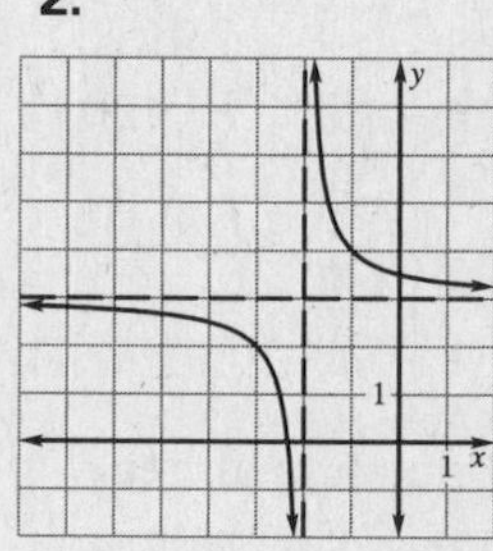

3.

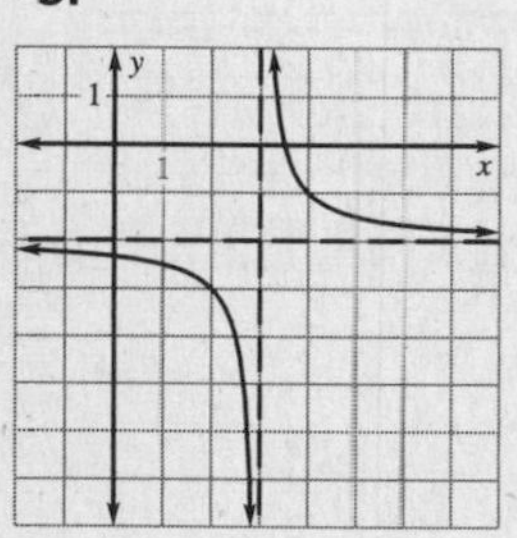

4.

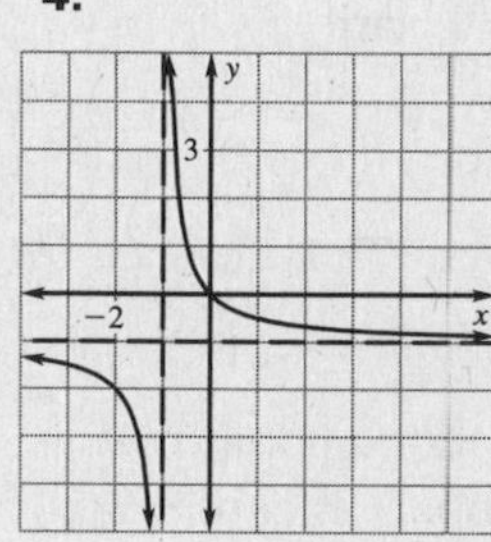

Practice A

1. all real numbers except 5 **2.** all real numbers except -6 **3.** all real numbers except 0 **4.** $x = -1$ **5.** $x = 2$ **6.** $x = -3$ **7.** $y = \frac{1}{2}$; all real numbers except $\frac{1}{2}$ **8.** $y = -1$; all real numbers except -1 **9.** $y = 6$; all real numbers except 6 **10.** B **11.** C **12.** A

13.

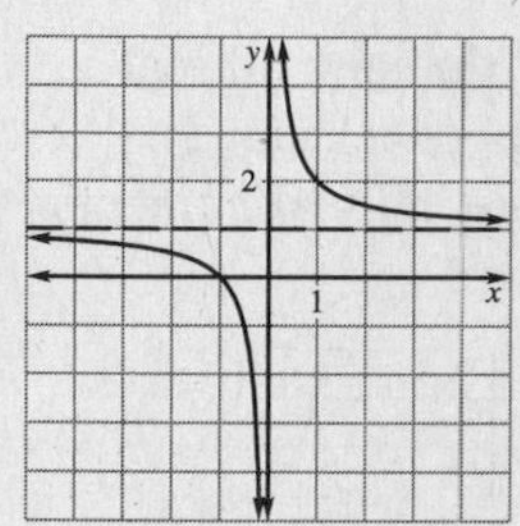

14.

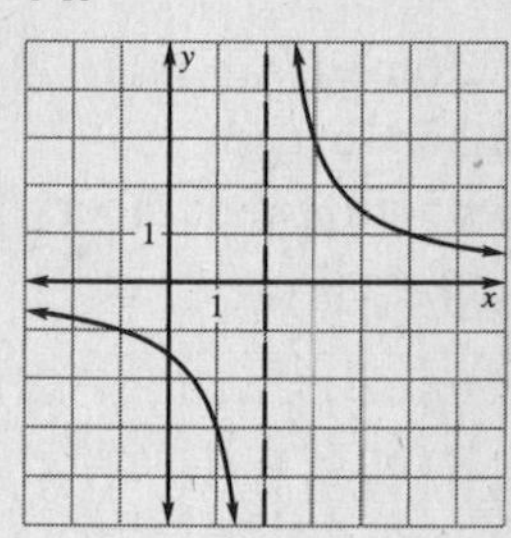

15.

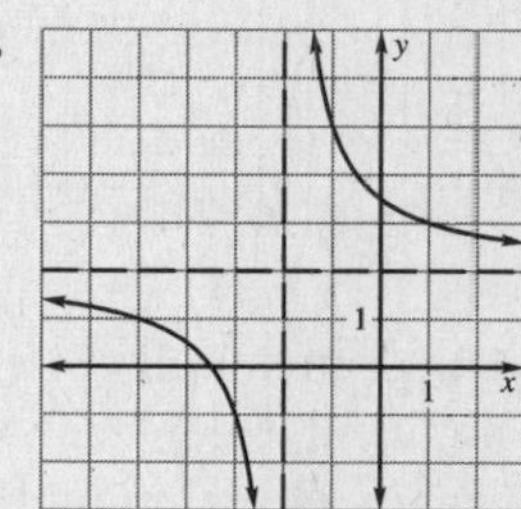

16. $C = 7x + 250$

17. $A = \dfrac{7x + 250}{x}$

18.

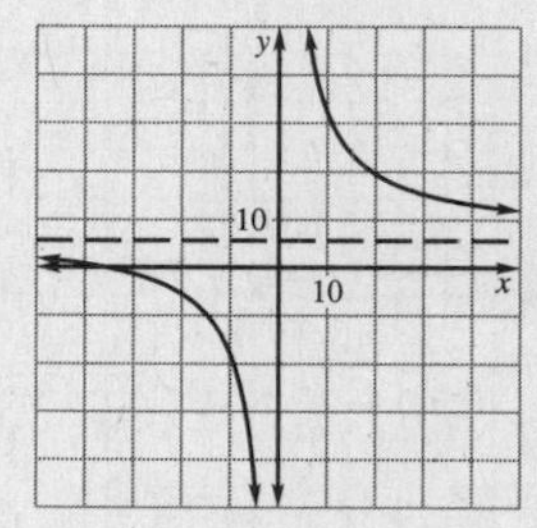

Practice B

1. $y = -5$; $x = -4$; domain: all real numbers except -4; range: all real numbers except -5

2. $y = \frac{3}{4}$; $x = -\frac{1}{4}$; domain: all real numbers except $-\frac{1}{4}$; range: all real numbers except $\frac{3}{4}$

3. $y = 2$; $x = \frac{2}{3}$; domain: all real numbers except $\frac{2}{3}$; range: all real numbers except 2 **4.** B **5.** C

6. A

7. domain: all real numbers except 0; range: all real numbers except 0

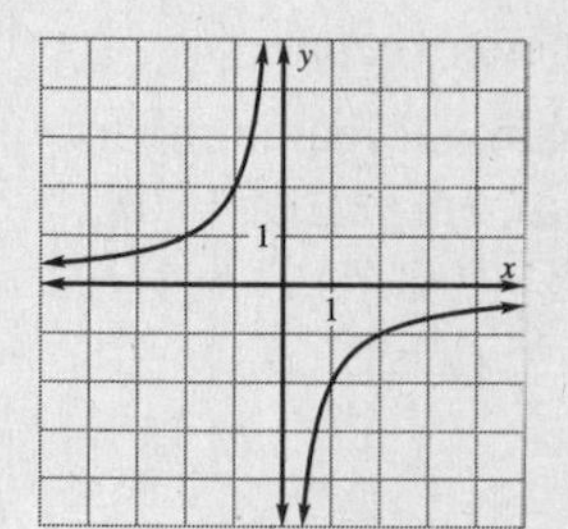

8. domain: all real numbers except 2; range: all real numbers except 3

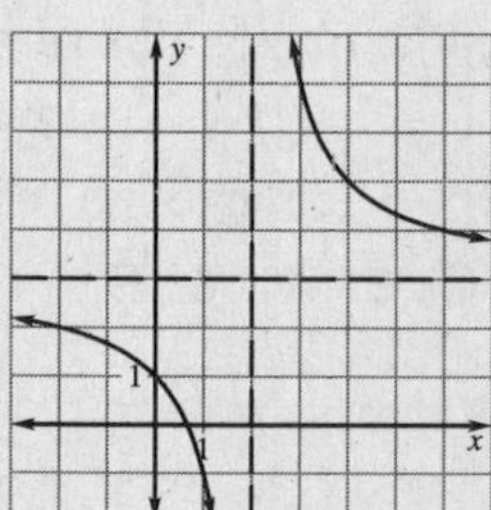

Lesson 9.2 *continued*

9. domain: all real numbers except -3; range: all real numbers except -1

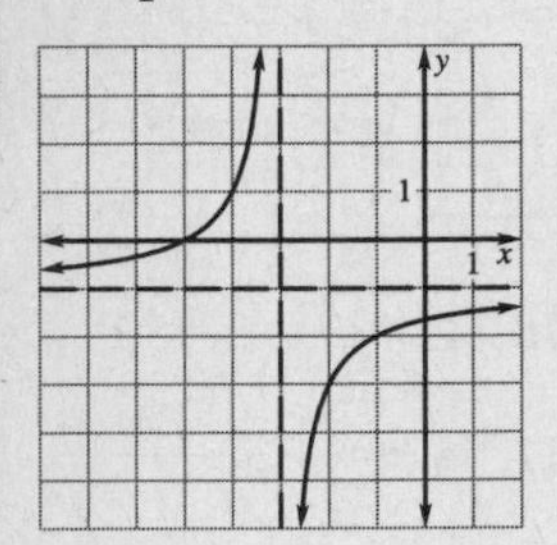

10. domain: all real numbers except 3; range: all real numbers except 1

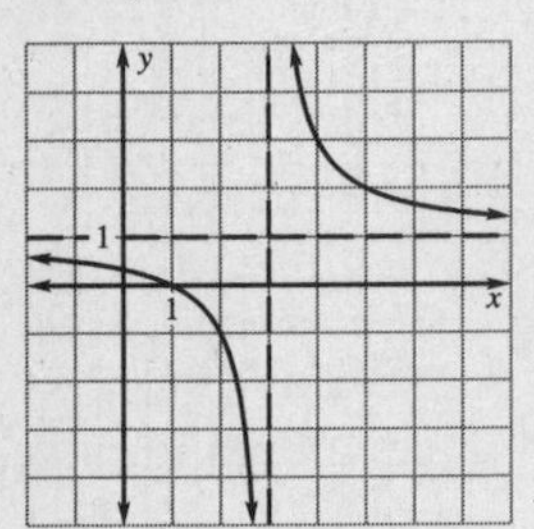

11. domain: all real numbers except $\frac{3}{2}$; range: all real numbers except $-\frac{3}{2}$

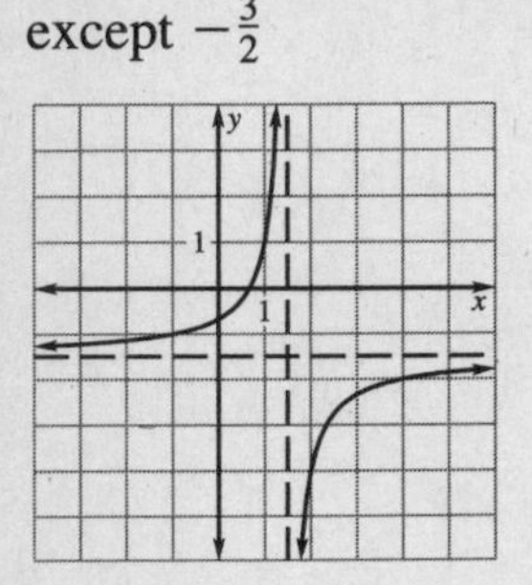

12. domain: all real numbers except $\frac{1}{2}$; range: all real numbers except $\frac{1}{2}$

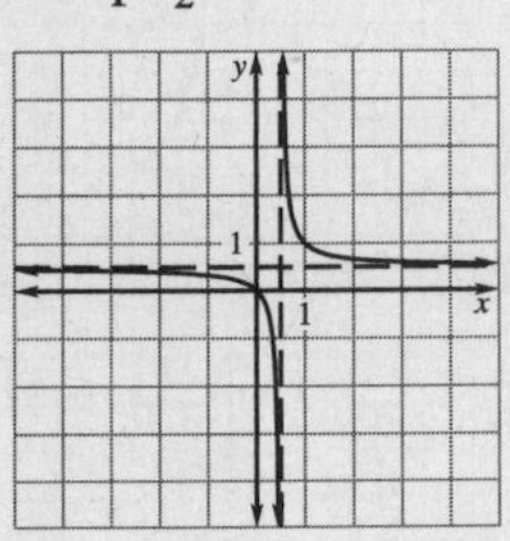

13.

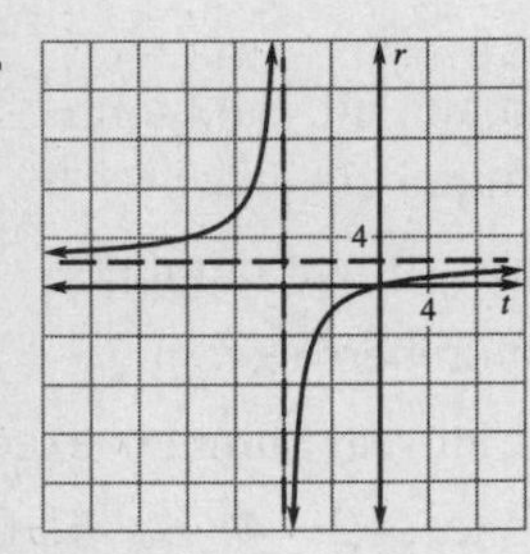

14. $y = 2$; the amount of rain will be less than 2 inches

15. 0.8 inch

Practice C

1. $y = 5$; $x = \frac{1}{2}$; domain: all real numbers except $\frac{1}{2}$; range: all real numbers except 5

2. $y = -\frac{3}{4}$; $x = \frac{1}{8}$; domain: all real numbers except $\frac{1}{8}$; range: all real numbers except $-\frac{3}{4}$

3. $y = -10$; $x = -7$; domain: all real numbers except -7; range: all real numbers except -10

4. C **5.** A **6.** B

7. domain: all real numbers except 0; range: all real numbers except -1

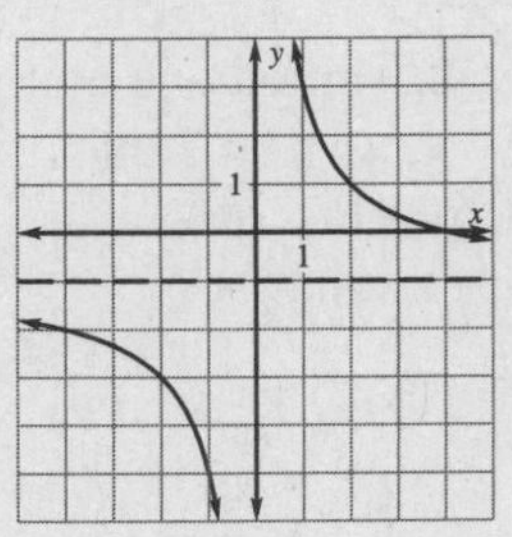

8. domain: all real numbers except 3; range: all real numbers except 4

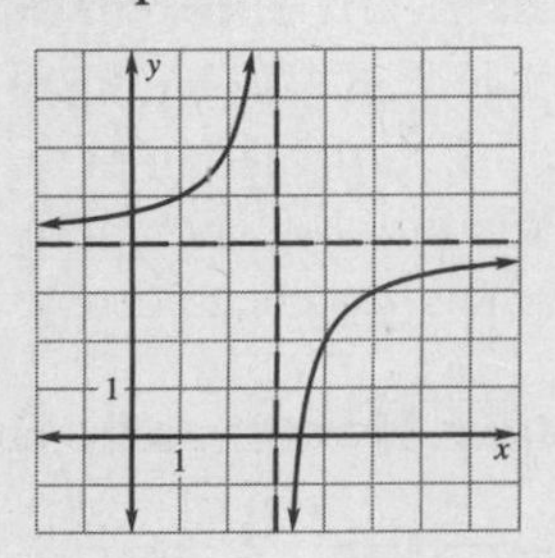

9. domain: all real numbers except $-\frac{1}{4}$; range: all real numbers except -3

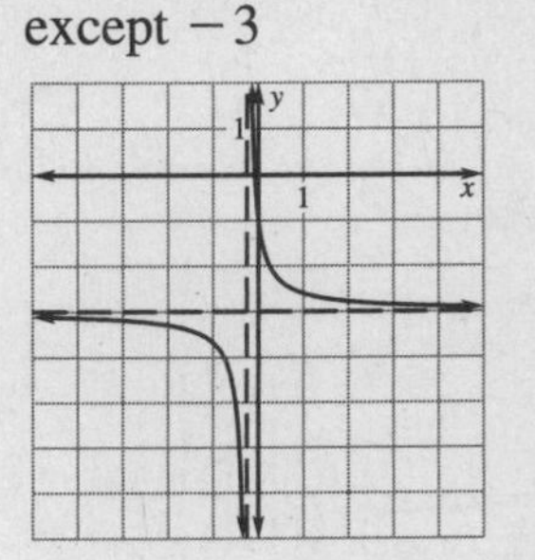

10. domain: all real numbers except $\frac{3}{2}$; range: all real numbers except 2

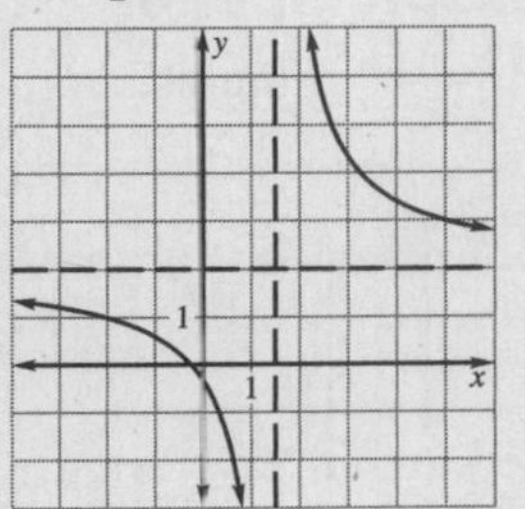

11. domain: all real numbers except $-\frac{2}{3}$; range:all real numbers except $\frac{1}{3}$

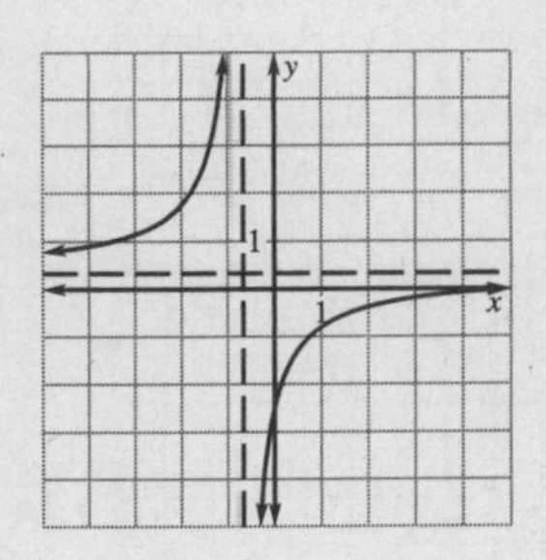

12. domain: all real numbers except -3; range:all real numbers except -5

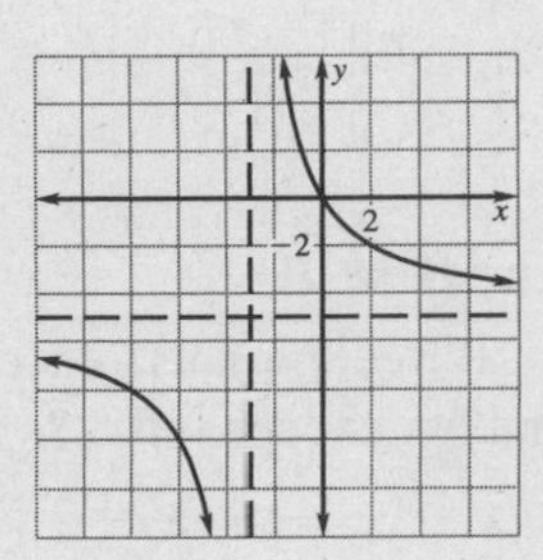

13.

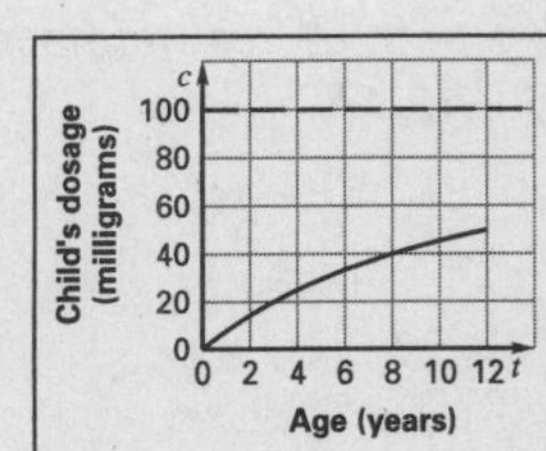

14. 40 milligrams

15. $y = 100$; the child's dose will be less than 100 milligrams

Lesson 9.2 *continued*

Reteaching with Practice

1. domain: all real numbers except 0
range: all real numbers except 0

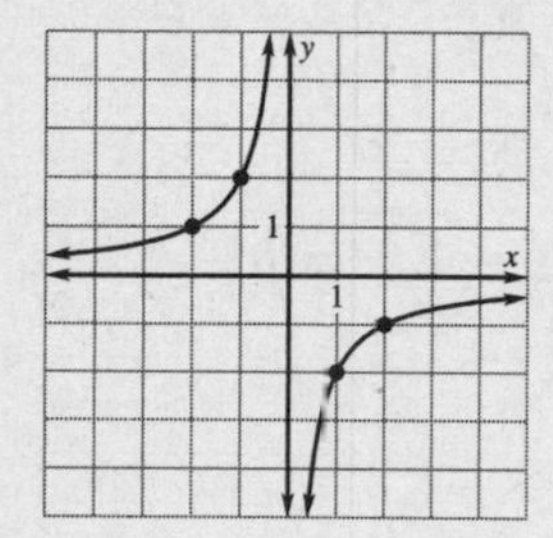

2. domain: all real numbers except 0
range: all real numbers except 5

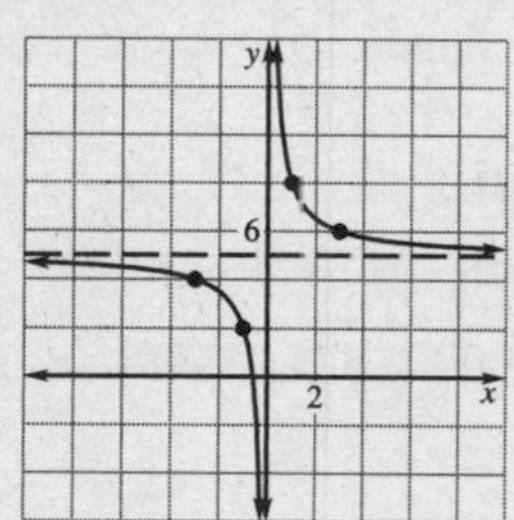

3. domain: all real numbers except 3
range: all real numbers except 1

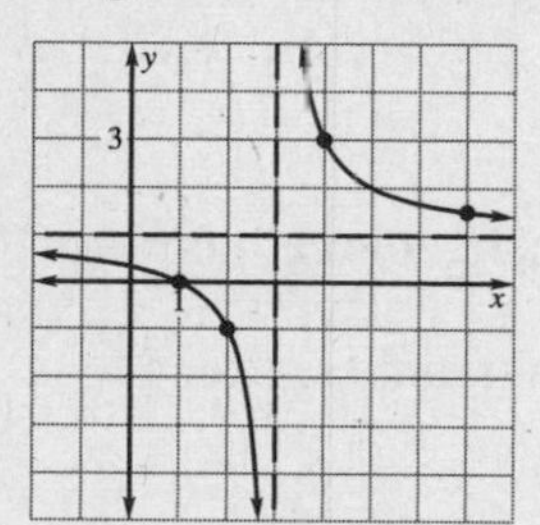

4. domain: all real numbers except -5
range: all real numbers except -2

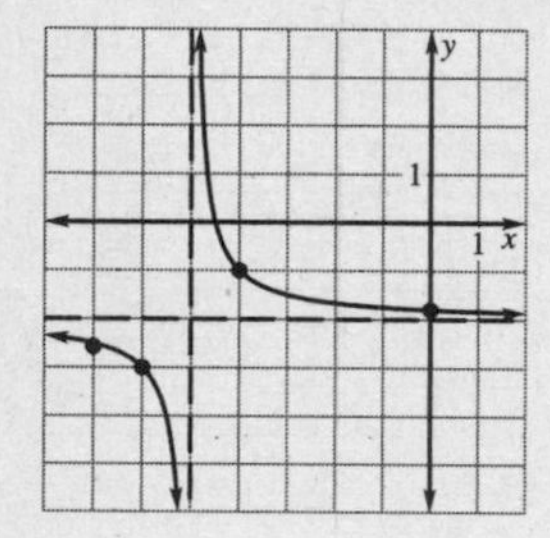

5. domain: all real numbers except -2
range: all real numbers except -1

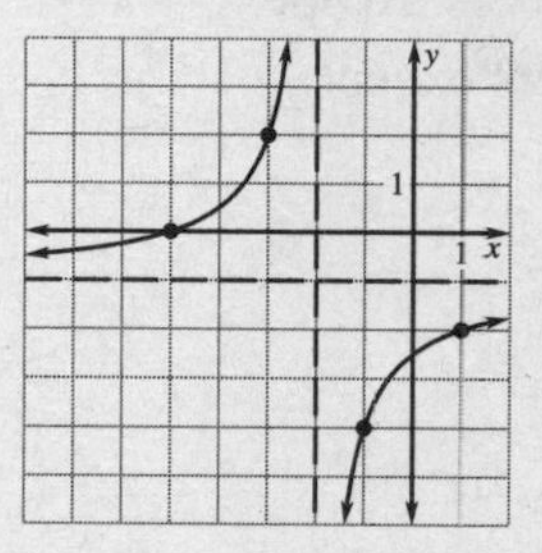

6. domain: all real numbers except 1
range: all real numbers except -4

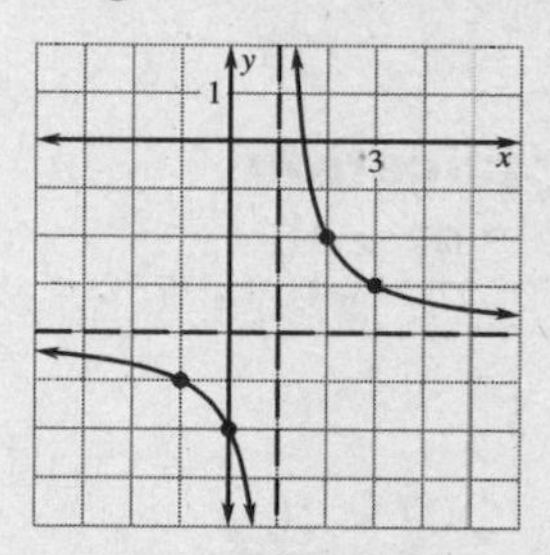

7. domain: all real numbers except -2
range: all real numbers except 1

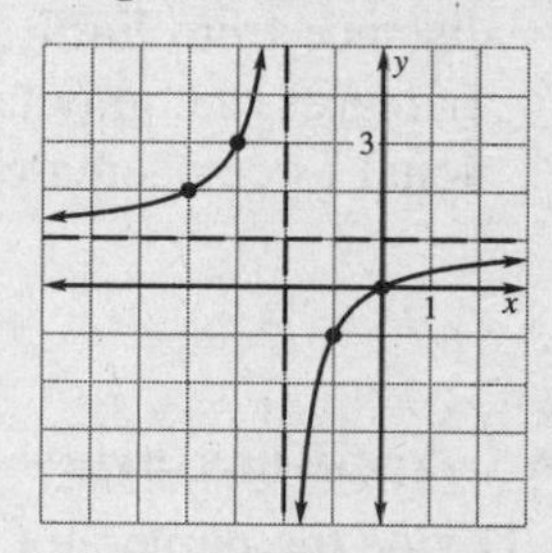

8. domain: all real numbers except 4
range: all real numbers except 2

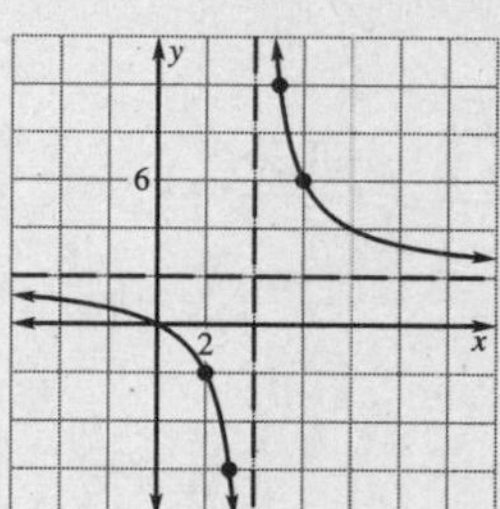

9. domain: all real numbers except -3
range: all real numbers except 1

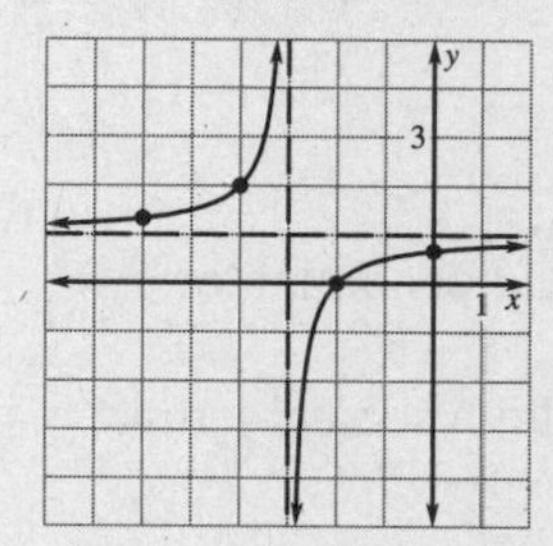

Lesson 9.2 *continued*

10. domain: all real numbers except $\frac{3}{2}$
range: all real numbers except $\frac{1}{2}$

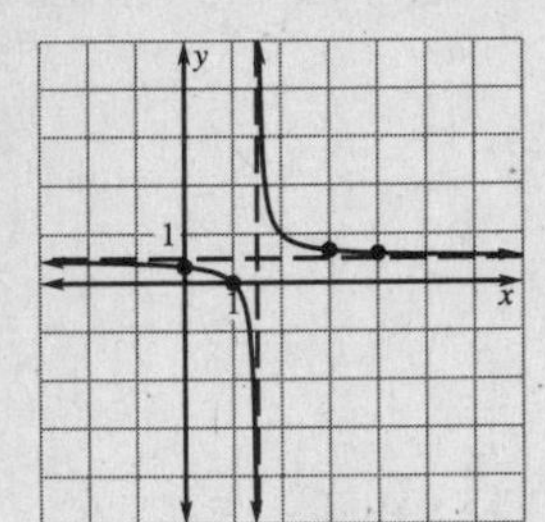

11. about \$32

Interdisciplinary Application

1. 1.98×10^{29} **2.** $F = \dfrac{2.07 \times 10^{21}}{x^2}$, where x is the distance from Earth.

3. $0 < x \leq 25{,}000$, $F \geq \dfrac{2.07 \times 10^{21}}{(25{,}000)^2}$

4. As the satellite's distance from Earth increases, the gravitational force decreases.

Challenge: Skills and Applications

1. a. a straight line **b.** Divide the numerator and denominator by c; then $a' = \dfrac{a}{c}$, $b' = \dfrac{b}{c}$, and $d' = \dfrac{d}{c}$.

2. $y = \dfrac{4}{x - 3} - 1$ **3.** $y = \dfrac{-12}{2x + 4} + 2$

4. $y = \dfrac{8}{3x - 2} + 1$

5. First write the equation as $y = \dfrac{\left(\frac{a}{c}\right)x + \frac{b}{c}}{x + \frac{d}{c}}$. Long division then shows that $y = \dfrac{p}{x - \left(-\frac{d}{c}\right)} + \dfrac{a}{c}$, where $p = \dfrac{bc - ad}{c^2}$.

6. Dividing numerator and denominator by x, you have $\dfrac{a + \frac{b}{x}}{c + \frac{d}{x}}$, and both second terms approach 0 as $x \to +\infty$, so the fraction approaches $\dfrac{a}{c}$.

7. a. A horizontal line: $y = -\frac{3}{4}$

b. $ad = bc$, or $ad - bc = 0$

8. a. $y = \dfrac{b - dx}{cx - a}$ **b.** $x = k$, $y = h$

Lesson 9.3

Warm-Up Exercises

1. 3, −3 **2.** 4, −1 **3.** 1, −1, 0 **4.** 5, −1
5. no real solutions

Daily Homework Quiz

1. $x = -1$, $y = 2$; domain: all real numbers except −1, range: all real numbers except 2

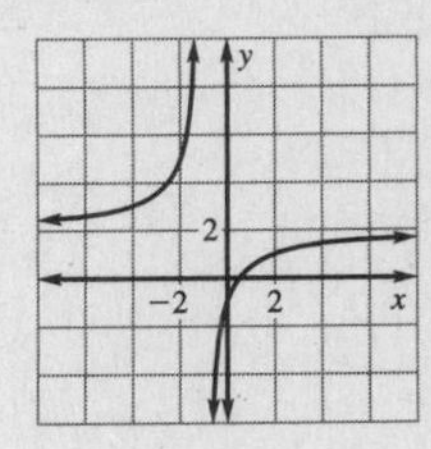

2. $x = \dfrac{-3}{2}$, $y = \dfrac{1}{2}$; domain: all real numbers except $\dfrac{-3}{2}$, range: all real numbers except $\dfrac{1}{2}$.

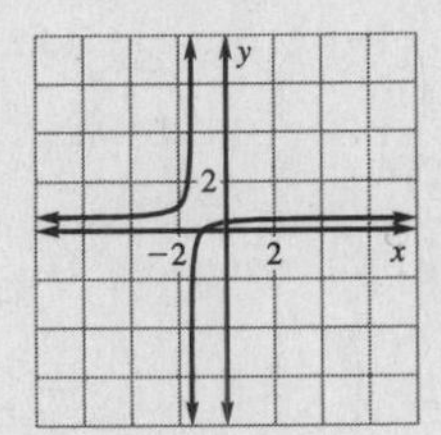

Lesson Opener

Allow 5 minutes.

1. $x = -1$, $x = 1$, $y = 1$

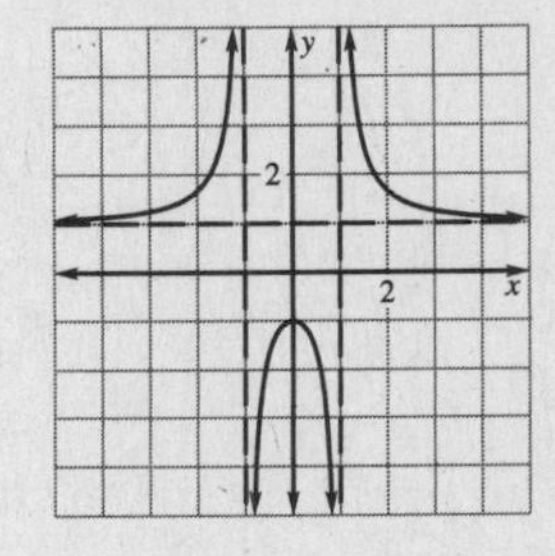

Lesson 9.3 *continued*

2. $y = 2$

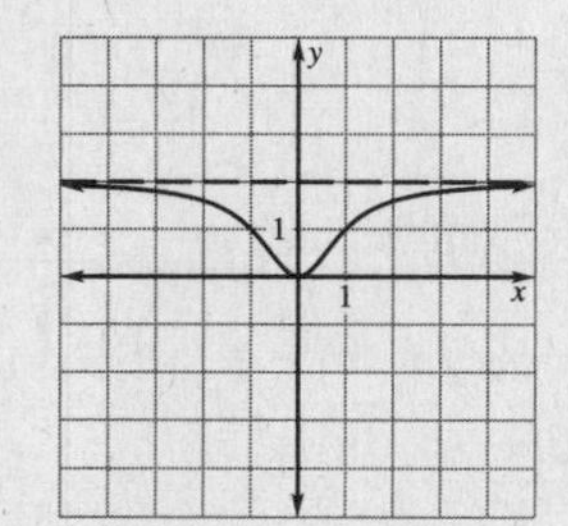

3. $x = -2, x = 1, y = 0$

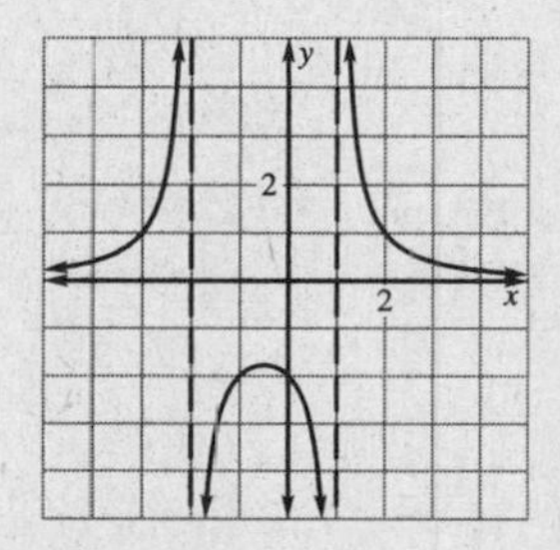

4. $x = -2$

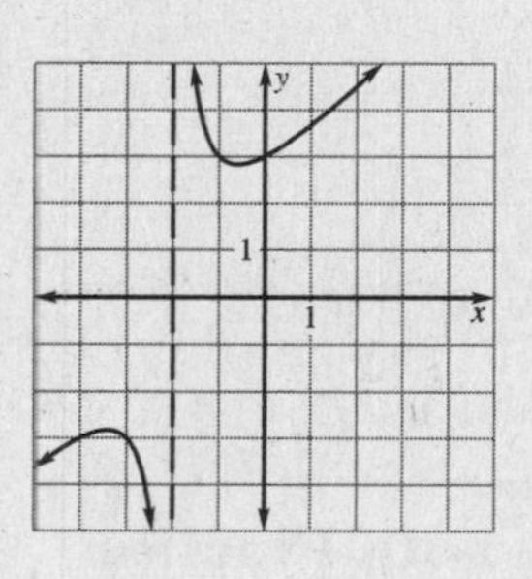

Practice A

1. x-intercept: 0; vertical asymptotes: $x = 5$, $x = -5$ **2.** x-intercept: -1; vertical asymptotes: $x = 3, x = -2$ **3.** x-intercept: $\frac{1}{2}$; vertical asymptote: $x = 0$ **4.** no x-intercepts; vertical asymptote: $x = -5$ **5.** x-intercepts: $9, -1$; no vertical asymptotes **6.** x-intercepts: $\sqrt{6}, -\sqrt{6}$; vertical asymptote: $x = 0$ **7.** B **8.** A **9.** C

10.

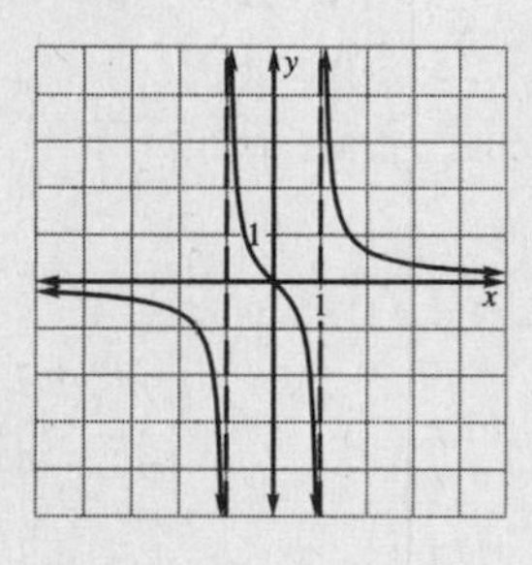

11.

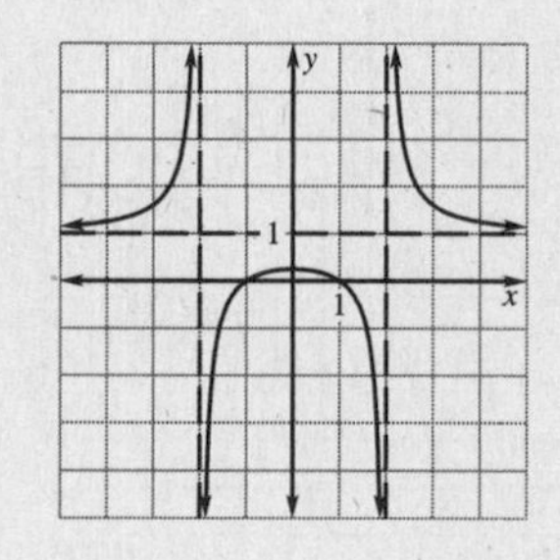

12.

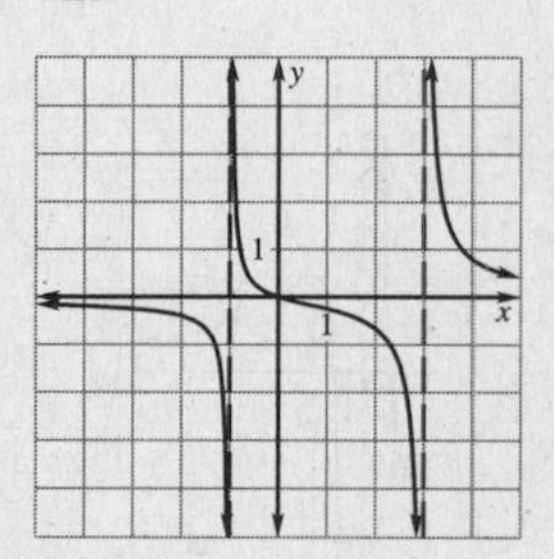

13.

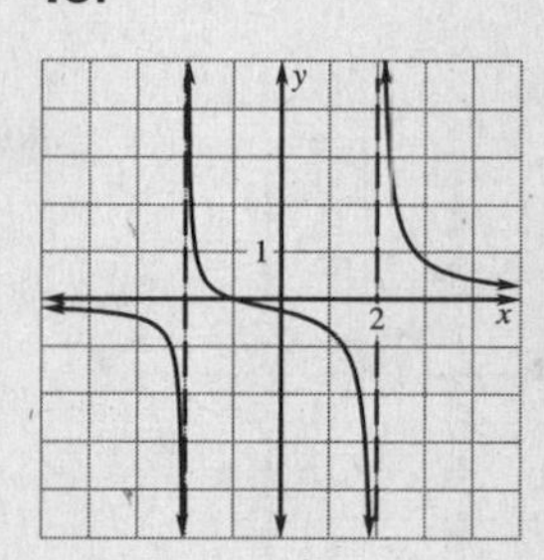

14.

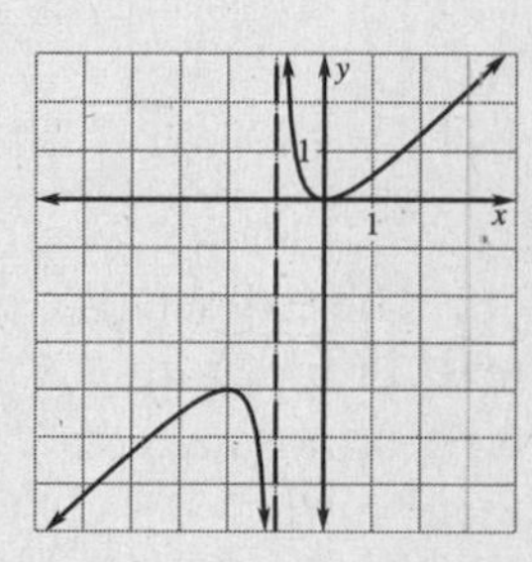

15.

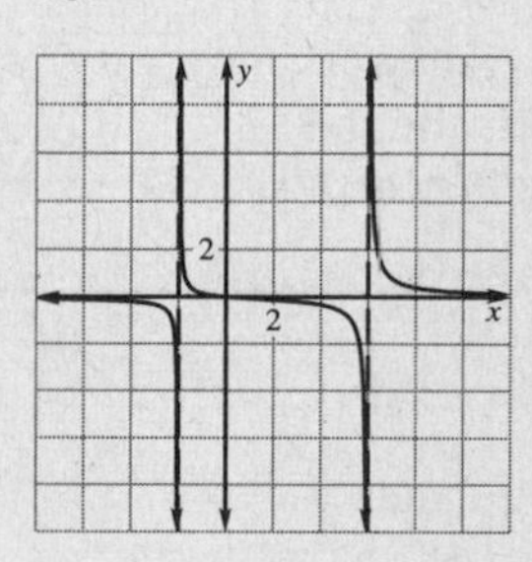

16. 15 ft by 30 ft

Practice B

1. x-intercepts: $-\frac{1}{2}, 4$; vertical asymptote: $x = -5$ **2.** no x-intercepts; vertical asymptotes: $x = 1, x = -1$ **3.** x-intercepts: 0; vertical asymptotes: $x = -4$ **4.** B **5.** C **6.** A

7.

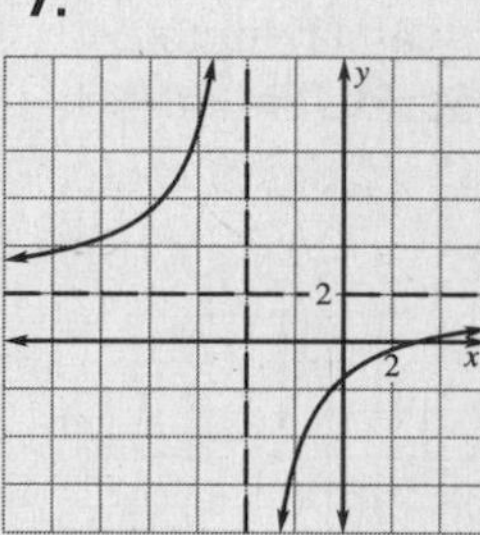

8.

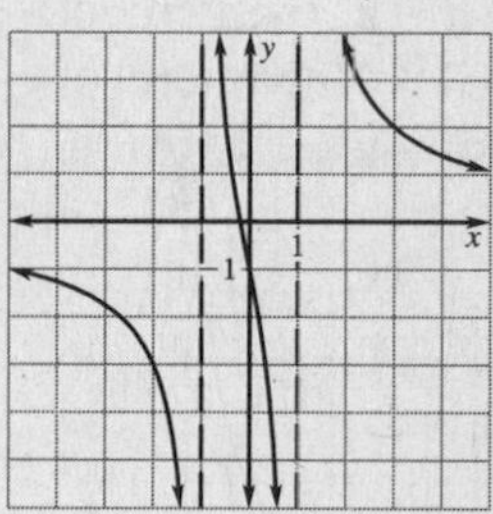

9.

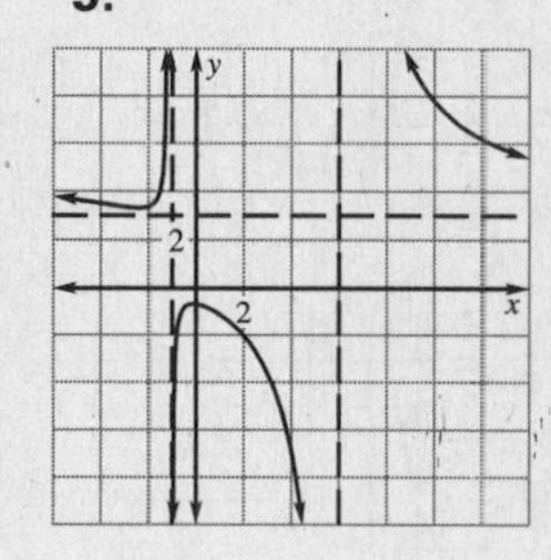

10.

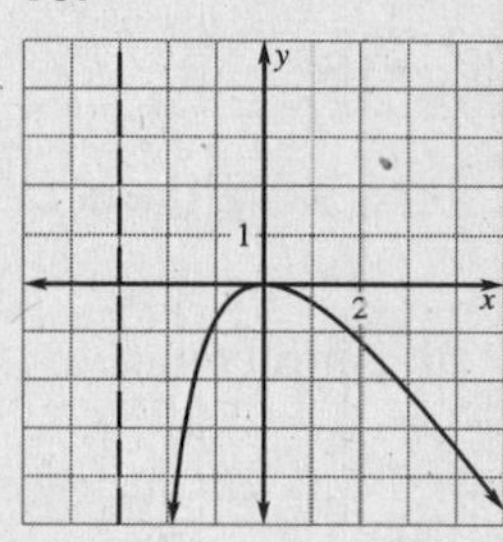

Lesson 9.3 *continued*

11.

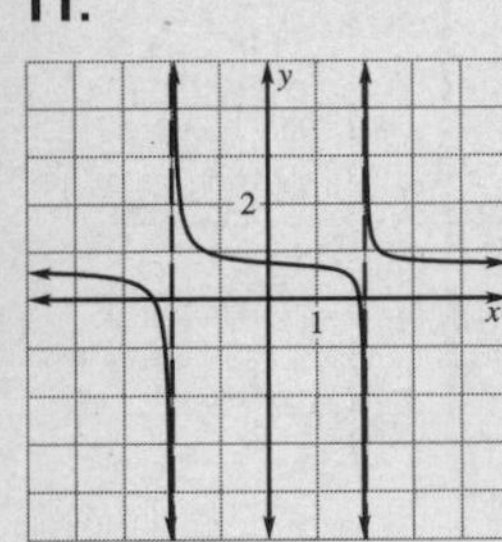

12.

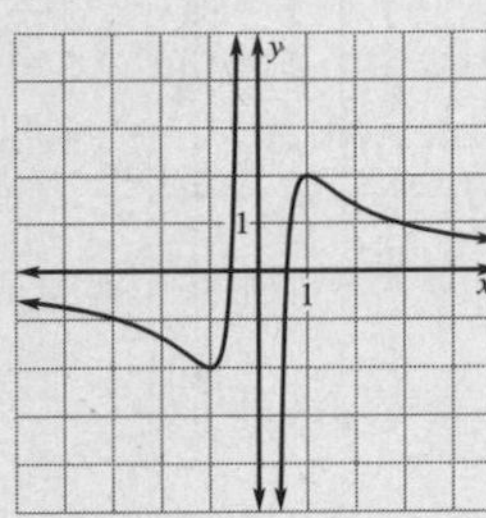

13. Answers may vary.

Sample answer: $y = \dfrac{x - 5}{x^2 + 3x}$

14.

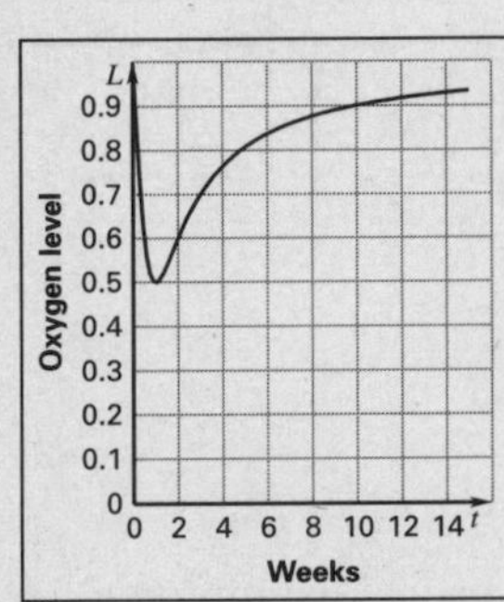

15. The oxygen level dropped to 50% of normal, then slowly increased to 93% of normal.

Practice C

1. no x-intercepts; vertical asymptotes: $x = -\frac{2}{3}$, $x = 3$ **2.** x-intercept: -2; vertical asymptote $x = 0$ **3.** x-intercepts: -3, -5; no vertical asymptotes **4.** A **5.** C **6.** B

7.

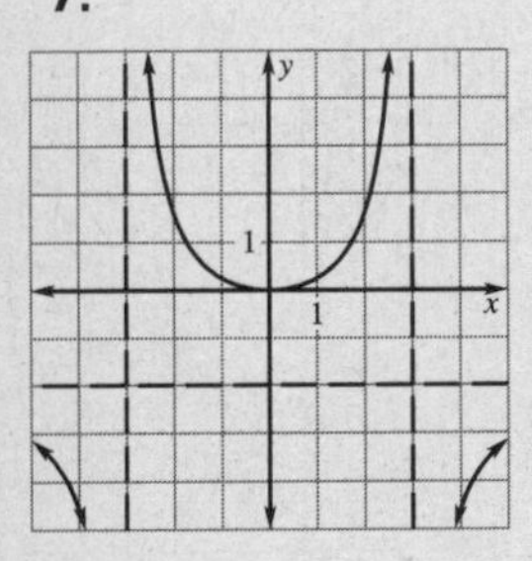

8.

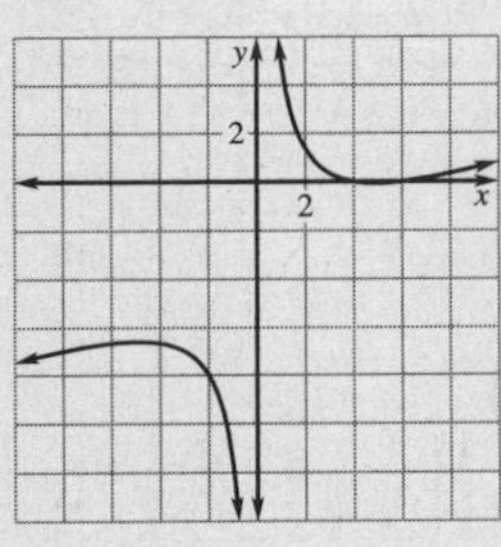

9.

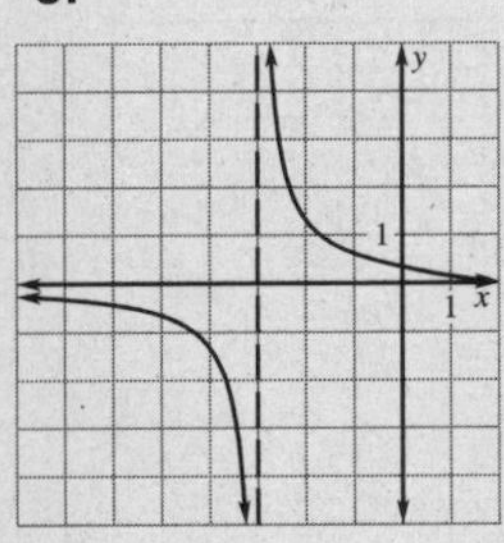

10.

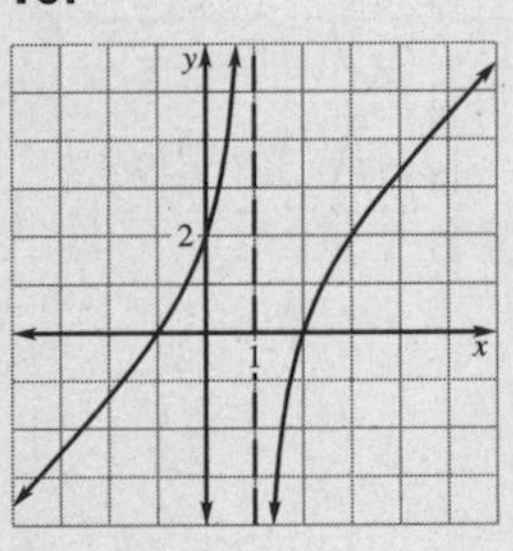

11.

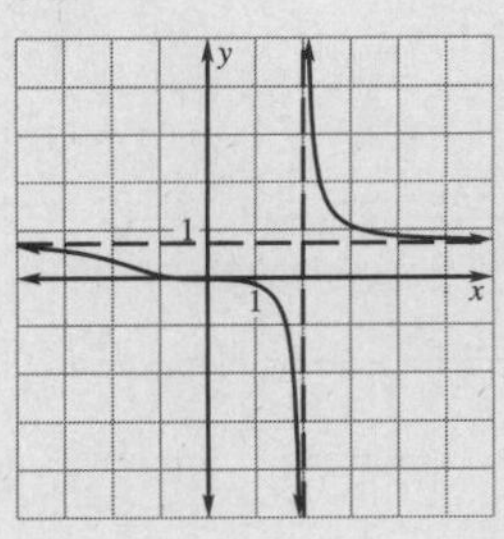

12.

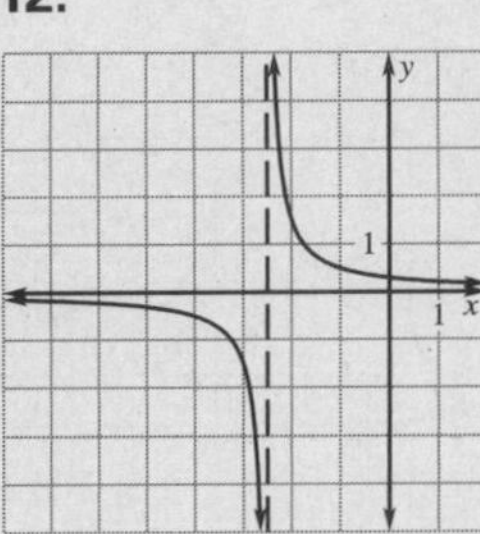

13. Answers may vary.

Sample answer: $y = \dfrac{2x^2}{x^2 - x - 12}$

14. $h = \dfrac{300}{\pi r^2}$

15. $S = 2\pi r^2 + \dfrac{600}{r}$

16.

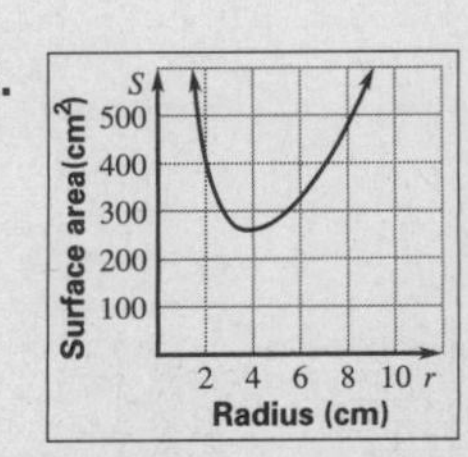

17. $r \approx 3.67$ cm and $h \approx 8.42$ cm

Reteaching with Practice

1.

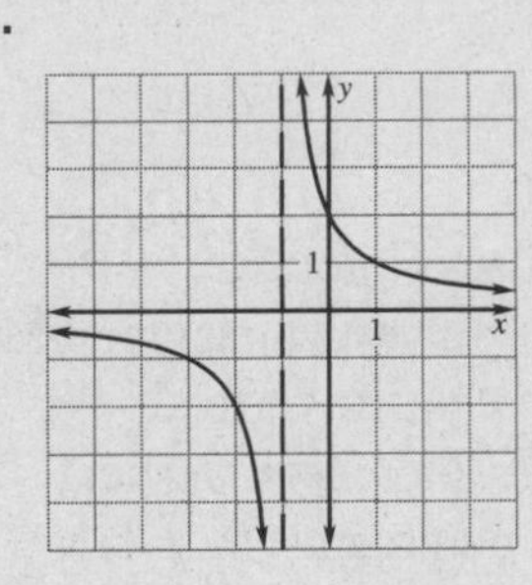

2.

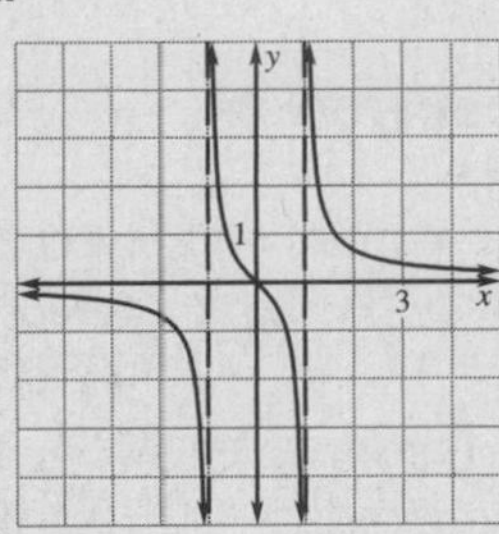

3.

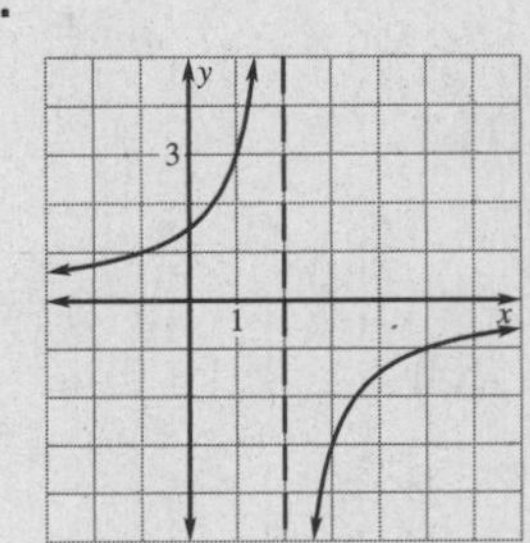

4.

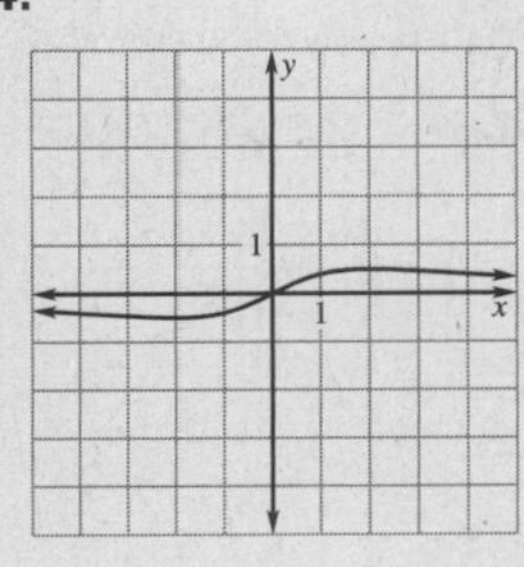

5.

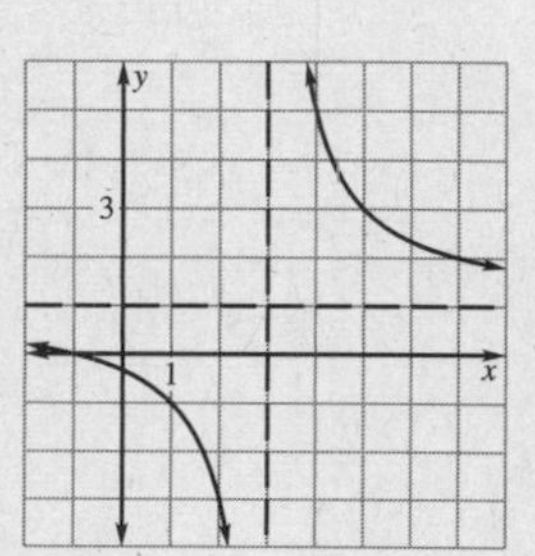

6.

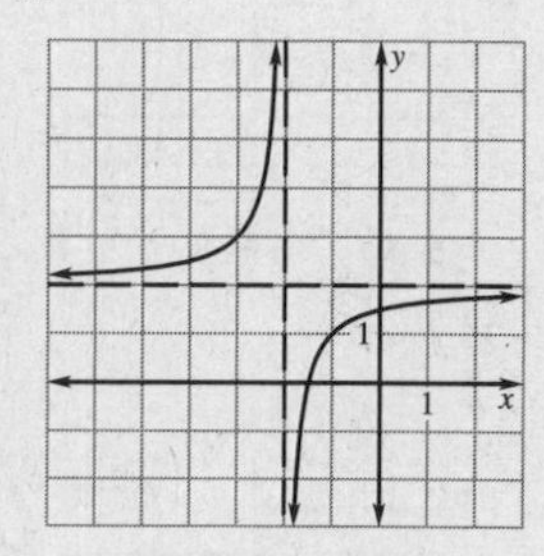

7.

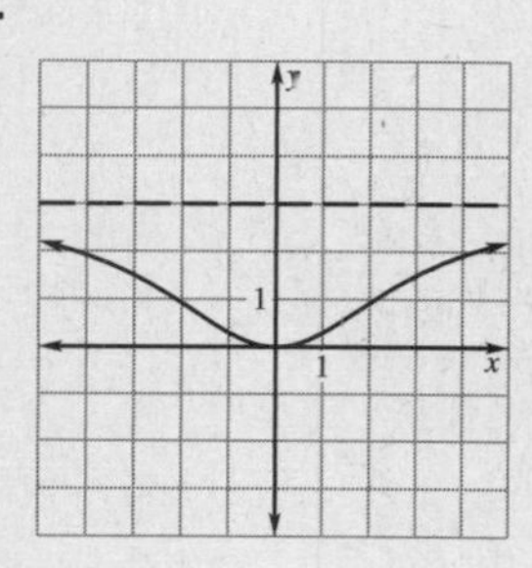

8.

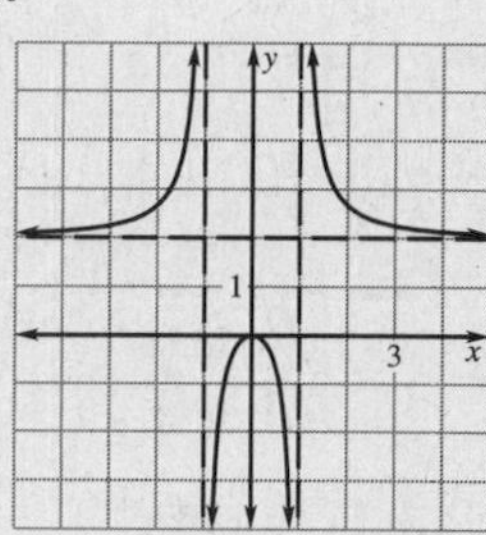

9.

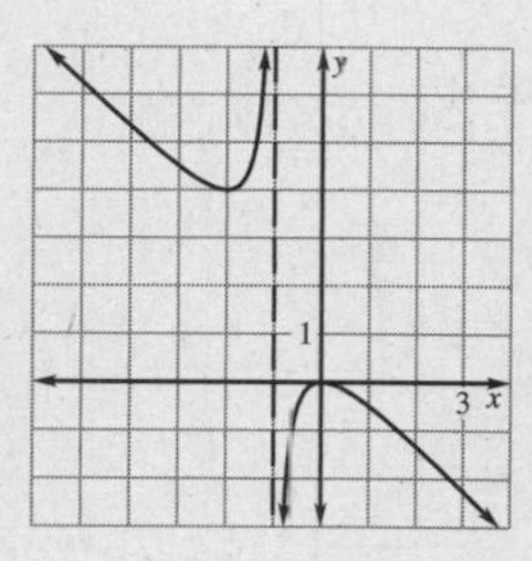

10.

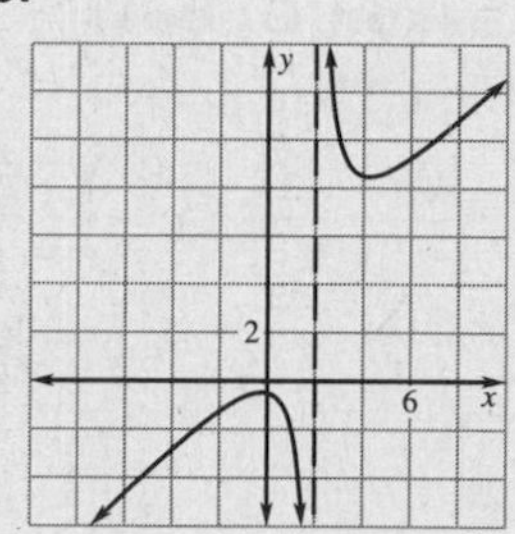

11.

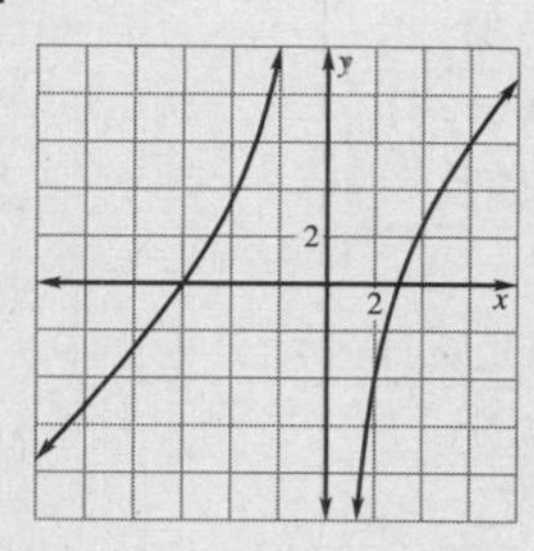

12.

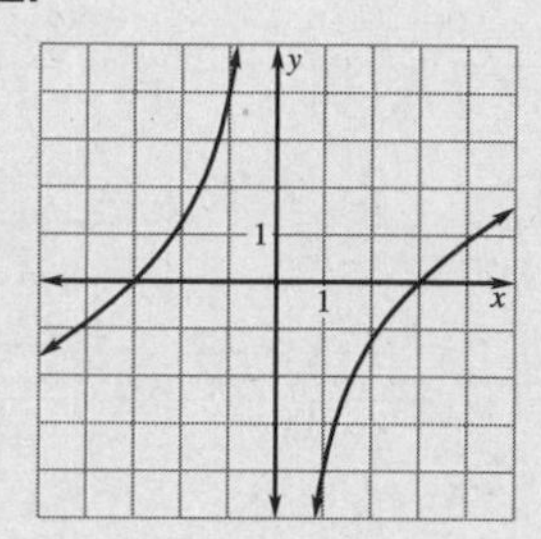

Real-Life Application

1.

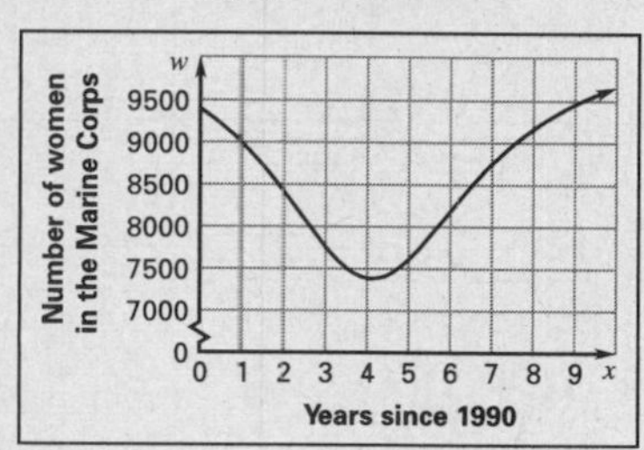

2. 1992: 8414; 1994: 7390; 1996: 8188

3. no x-intercept

4. no vertical asymptote; $y = 10{,}351.25$ (horizontal asymptote) **5.** 10,084 **6.** *Sample answer:* Increased number of job and educational opportunities would cause more women to join the Marine Corps.

Challenge: Skills and Applications

1. *Sample answer:* $y = \dfrac{-2x^2}{x^2 - 9}$

2. *Sample answer:* $y = \dfrac{x - 1}{x^2 - 4x}$

3. *Sample answer:* $y = \dfrac{x^2}{x^2 + 1}$

4. a.

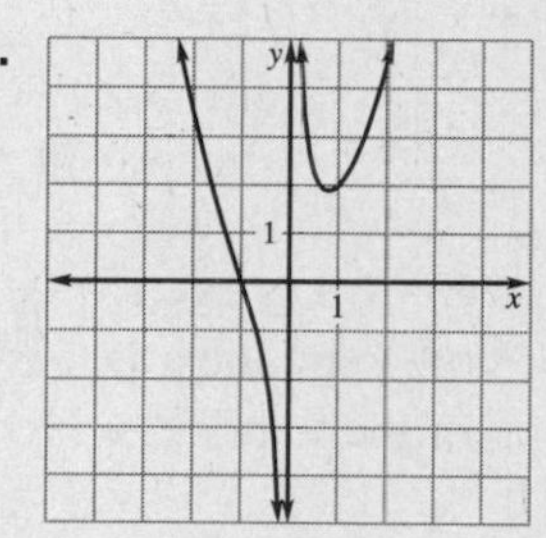

b.

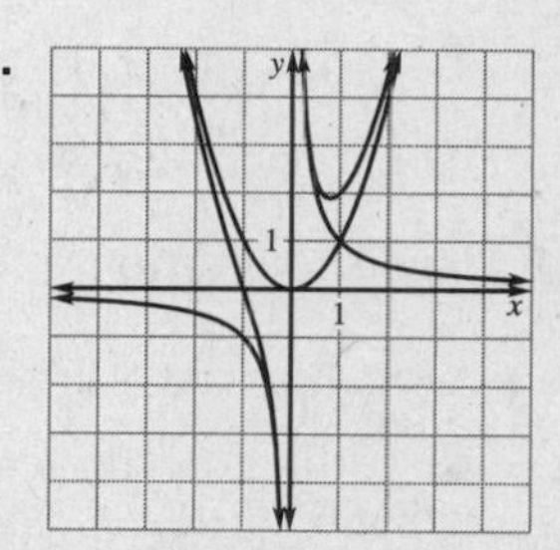

The original graph gets closer and closer to the graph of $y = \dfrac{1}{x}$ as $x \to 0$, and gets closer and closer to the graph of $y = x^2$ as $x \to +\infty$ and as $x \to -\infty$.

5. a.

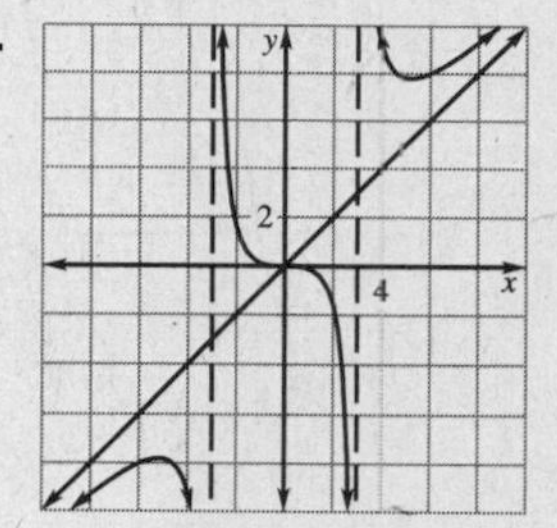

b. The graph of $y = x$ is an asymptote for the graph of the rational function.

Lesson 9.3 *continued*

6. a.

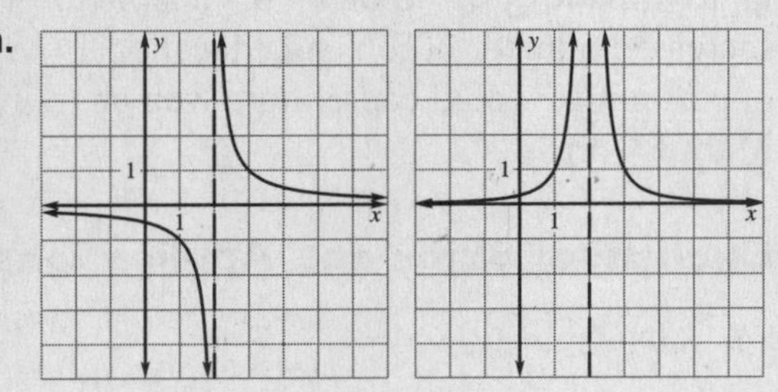

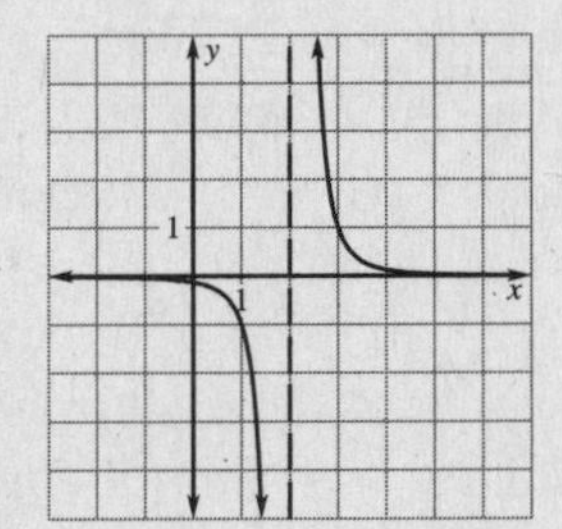

b. Near the asymptote, if n is even $y \to +\infty$ or $y \to -\infty$ for both branches; if n is odd, $y \to +\infty$ for one branch and $y \to -\infty$ for the other branch.

7. $A = -1, B = 4$

Quiz 1

1. direct variation **2.** inverse variation

3. $y = \frac{-15}{x}$; 3 **4.** $x = \frac{-1}{6}yz$; -48

5.

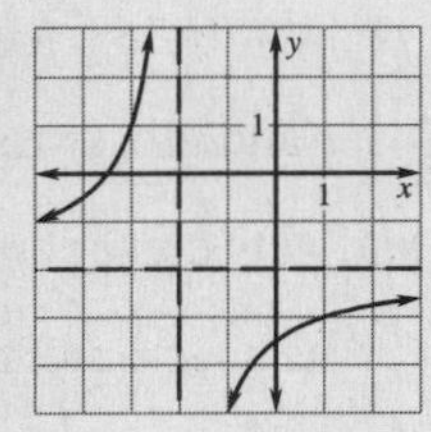

6.

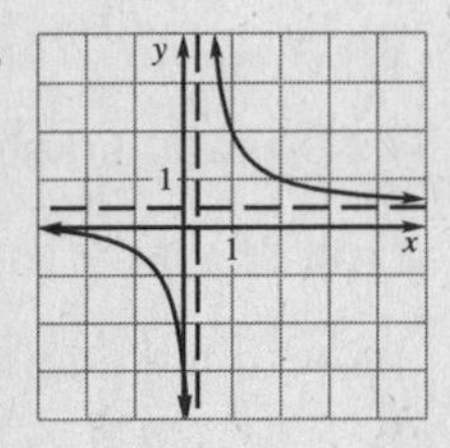

7.

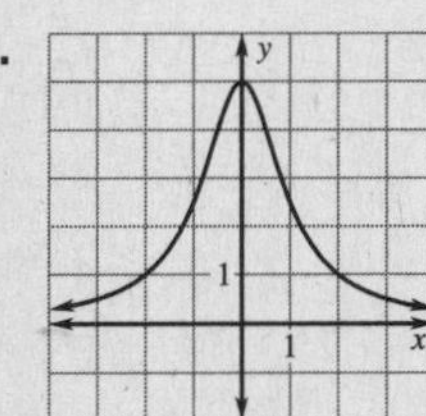

8.

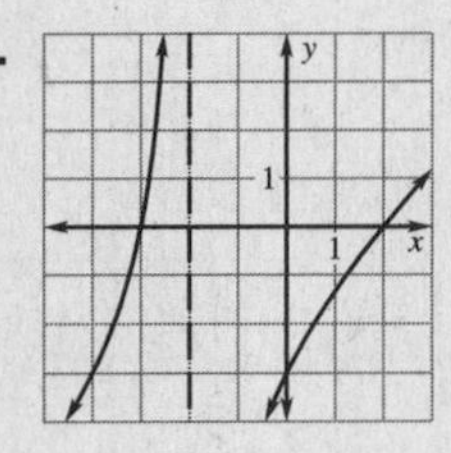

9. \$4.97

Lesson 9.4

Warm-Up Exercises

1. $(x - 1)(x + 4)$ **2.** $(x + 2)(x + 3)$

3. $(2x - 3)(2x + 3)$ **4.** $x(6x + 1)$

5. $(2x + 1)(4x^2 - 2x + 1)$

Daily Homework Quiz

1.

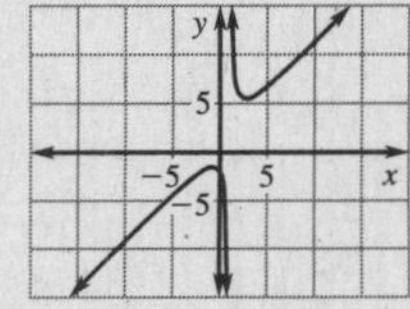

$x = 1$

2.

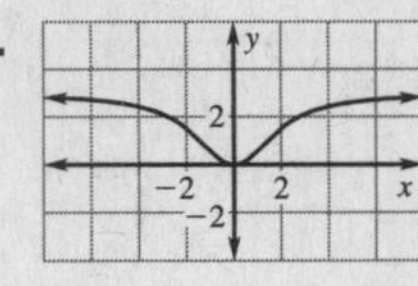

$y = 3$

3.

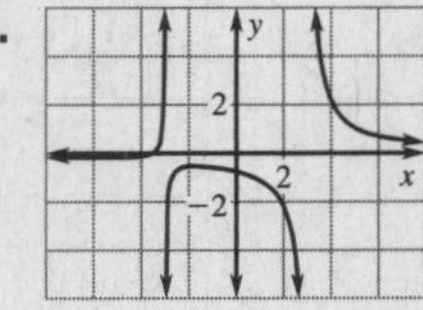

$y = 0, x = \pm 3$

Lesson Opener

Allow 20 minutes.

$$\frac{10x - 5}{x^2 - 7x + 12} = \frac{5(2x - 1)}{(x - 4)(x - 3)}$$

$$\frac{x^2 - 5x + 4}{6x - 4} = \frac{(x - 4)(x - 1)}{2(3x - 2)}$$

$$\frac{3x^2 - x - 4}{6x - 2} = \frac{(x + 1)(3x - 4)}{2(3x - 1)}$$

$$\frac{3x + 9}{x^2 + 3x - 4} = \frac{3(x + 3)}{(x - 1)(x + 4)}$$

$$\frac{2x^2 + 9x + 9}{15x - 5} = \frac{(x + 3)(2x + 3)}{5(3x - 1)}$$

$$\frac{3x^2 - 11x - 4}{x^2 + 3x} = \frac{(x - 4)(3x + 1)}{x(x + 3)}$$

$$\frac{x^2 + 6x + 8}{2x^2 - 5x + 2} = \frac{(x + 2)(x + 4)}{(x - 2)(2x - 1)}$$

$$\frac{4x^2 + 3x - 1}{x^2 - 4x} = \frac{(x + 1)(4x - 1)}{x(x - 4)}$$

$$\frac{2x^2 - 7x - 4}{4x^2 + 12x + 9} = \frac{(x - 4)(2x + 1)}{(2x + 3)^2}$$

$$\frac{3x^2 + 8x + 4}{4x^2 - 4x - 3} = \frac{(x + 2)(3x + 2)}{(2x - 3)(2x + 1)}$$

$$\frac{3x^2 + 11x + 6}{3x^2 + x - 4} = \frac{(x + 3)(3x + 2)}{(x - 1)(3x + 4)}$$

$$\frac{4x + 6}{4x^2 - 11x - 3} = \frac{2(2x + 3)}{(x - 3)(4x + 1)}$$

Lesson 9.4 *continued*

Practice A

1. $\frac{4x}{2x+3}$ 2. $\frac{x+3}{x+1}$ 3. $\frac{x-4}{x-3}$ 4. $\frac{x-6}{x+5}$

5. not possible 6. $\frac{x-1}{x+1}$ 7. $\frac{8x}{5}$ 8. $\frac{3}{4x^2}$

9. $\frac{x-2}{x+1}$ 10. $\frac{x(x+3)}{x+1}$ 11. $\frac{x^3}{2}$ 12. $\frac{20x}{3y^3}$

13. $\frac{x}{3(x-1)}$ 14. $\frac{x-11}{x+8}$ 15. $\frac{9x^5y}{2}$ 16. $\frac{1}{2x^2}$

17. $(x+4)(x+3)$ 18. $\frac{2(x+2)}{x+6}$ 19. $\frac{4(x+1)}{3(x-2)}$

Practice B

1. $\frac{x-9}{x-1}$ 2. $\frac{1}{x+2}$ 3. not possible

4. $\frac{1}{5x^5y^2}$ 5. $\frac{9x^6}{4y^5}$ 6. $\frac{6(x-2)}{x+3}$ 7. $\frac{3}{10}$ 8. $\frac{8x}{5}$

9. $\frac{x(x-2)}{5}$ 10. $\frac{(x-2)(x+6)}{5x^2}$ 11. $x+3$

12. $(x-5)(x+4)$ 13. $\frac{2(x+4)}{3x}$ 14. $\frac{8}{x}$

15. $\frac{y^2}{18x^2}$ 16. $\frac{7}{6x(x+5)}$ 17. $\frac{x}{3(x+3)}$

Practice C

1. $\frac{3x+1}{x+2}$ 2. not possible 3. $\frac{x+5}{x^2+5x+25}$

4. $x-6$ 5. $\frac{x-2}{x+1}$ 6. $\frac{3}{10}$ 7. $\frac{3x^{11}y}{25}$

8. $3(x-2)$ 9. $\frac{7(x-3)}{x(x+5)}$ 10. $\frac{x(x^2+2x+4)}{4(x+1)}$

11. $\frac{x+6}{4(x+2)}$ 12. $\frac{6}{5}$ 13. $\frac{5(x+5)}{4x^3}$

14. $\frac{(x+10)(x+2)}{(x+8)(x+1)}$

15. $\frac{1}{x^2(x+3)(x-3)(x+4)}$ 16. $\frac{\pi}{1}$

17. about 60,769 gallons

Reteaching with Practice

1. $\frac{y+9}{2}$ 2. $\frac{1}{2}$ 3. $\frac{1}{x+3}$ 4. $\frac{y}{y-1}$

5. $\frac{x(x+3)}{x+1}$ 6. 2 7. -1 8. $\frac{20x}{3y^3}$

9. $\frac{x}{3(x-1)}$

Interdisciplinary Application

1. Percent of trumpet players =
$$\frac{-953.05y^2 + 22{,}360.5y + 55{,}000}{-2355.64y^2 + 120{,}167.27y + 638{,}700}$$

2. about 11% 3. 5, 6, 7 4. about 320 members, about 37 trumpet players 5. about 11.4%

Challenge: Skills and Applications

1. a. $\frac{(x-3)(x-5)}{(x+1)(x-2)}$; $x=-1, x=2$

b. $x=a$ is a vertical asymptote of the product if and only if it is a vertical asymptote of one of the factors and is not a zero of either factor.

2. $(x-3)(x+4)$ 3. $\frac{1}{x^2+5x+25}$

4. a. $(x+a)^3 = x^3 + 3ax^2 + 3a^2x + a^3$

b. (i) $x-2$ (ii) $\frac{x+3}{x(x+2)}$

5. a. $(x-a)(x^{n-1} + ax^{n-2} + \cdots + a^{n-2}x + a^{n-1}) = x(x^{n-1} + ax^{n-2} + \cdots + a^{n-2}x + a^{n-1}) - a(x^{n-1} + ax^{n-2} + \cdots + (a^{n-2}x + a^{n-1}) = (x^n + ax^{n-1} + \cdots + (a^{n-2}x^2 + a^{n-1}x) - (ax^{n-1} + a^2x^{n-2} + \cdots + a^{n-1}x + a^n) = x^n - a^n$

b. (i) $\frac{1}{x+5}$ (ii) $\frac{1}{(x-2)^2}$

Lesson 9.5

Warm-Up Exercises

1. $\frac{4}{5}$ 2. $\frac{5}{4}$ 3. $\frac{1}{6}$ 4. $\frac{2}{3}$ 5. $\frac{6}{5}$

Daily Homework Quiz

1. a. $\frac{x-3}{x+4}$ b. $\frac{x-4}{x+4}$

2. a. $\frac{x-1}{3x^2(x+2)}$ b. $\frac{x+2}{x-1}$

Lesson 9.5 *continued*

c. $\frac{6x+1}{7}$ **d.** 1

Lesson Opener

Allow 10 minutes.

1. Because $\mathbf{x-1}$ and $\mathbf{x+3}$ have no common factors, the LCD is $\mathbf{(x-1)(x+3)}$.

2. Rewrite each expression using the LCD. Then add the numerators, and simplify if possible.

$$\frac{2}{x-1}+\frac{5}{x+3}$$
$$=\frac{2}{x-1}\cdot\frac{\mathbf{x+3}}{\mathbf{x+3}}+\frac{5}{x+3}\cdot\frac{\mathbf{x-1}}{\mathbf{x-1}}$$
$$=\frac{\mathbf{2(x+3)}}{\mathbf{(x-1)(x+3)}}+\frac{\mathbf{5(x-1)}}{\mathbf{(x-1)(x+3)}}$$
$$=\frac{\mathbf{7x+1}}{\mathbf{(x-1)(x+3)}}$$

Graphing Calculator Activity

1. $\frac{2}{(x+2)}+\frac{3}{(x-5)}=$
$\frac{2(x-5)+3(x+1)}{(x+2)(x-5)}=\frac{5x-7}{(x+2)(x-5)}$

3. $\frac{\left(\frac{2}{x}+3\right)(x)(x+2)}{\left(\frac{1}{(x+2)}+5\right)(x)(x+2)}=\frac{3x^2+8x+4}{5x^2+11x}$

Practice A

1. $\frac{4}{x}$ **2.** $\frac{5+x}{x+1}$ **3.** $\frac{2x-1}{x-3}$

4. $(x+1)(x-1)$ **5.** $(x+4)(x-4)$

6. $(x+2)(x-1)$ **7.** $2x(x+1)^2$

8. $\frac{15x+2}{3x^2}$ **9.** $\frac{x^2+4x-6}{2x^2}$

10. $\frac{-x-17}{(x+5)(x+1)}$ **11.** $\frac{6x+13}{x+3}$

12. $\frac{x+6}{(x+2)(x-2)}$ **13.** $\frac{x^2+4x}{(x+1)(x-1)}$

14. $\frac{2x-1}{x^2}$ **15.** $\frac{x+1}{x-1}$ **16.** $\frac{3x-x^2}{x-1}$

17. $R=\frac{R_1R_2}{R_1+R_2}$ **18.** $\frac{4}{3}$ ohms

Practice B

1. $(2x+1)(2x-1)$ **2.** $4(x+4)$

3. $x(x+1)(x-1)$ **4.** $\frac{7+x}{x-2}$

5. $\frac{2x-1}{(x+2)(x-1)}$ **6.** $\frac{6}{(x-6)(x+5)}$

7. $\frac{2(4x^2+5x-3)}{x^2(x+3)}$ **8.** $\frac{x}{x-1}$

9. $\frac{x^2+3x+9}{x(x+3)(x-3)}$ **10.** $\frac{4x+7}{x+3}$

11. $\frac{x^2+9x-12}{3x^2}$ **12.** $\frac{3x+1}{4x^2}$

13. $\frac{-11x+2}{3(x^2+x+2)}$ **14.** $\frac{1}{6}$

15. $I=\frac{-1374t^2-20{,}461t+1{,}627{,}410}{(85-t)(55-2t)}$

16. about 554,000 MDs; about 24,000 DOs

Practice C

1. $4(x+4)(x-4)$ **2.** $x(x+1)(x-1)^2$

3. $x(x-2)(x-6)$ **4.** $\frac{1}{x+1}$

5. $-\frac{2(x^2-8x-21)}{3(x-4)(x+3)}$ **6.** $\frac{5x^2+2}{x^3}$

7. $\frac{4(2x+5)}{x+2}$ **8.** $\frac{2x-1}{x}$ **9.** $\frac{-x-5}{(x+2)^2}$

10. $\frac{2(2x^2+4x-3)}{(x+3)^2(x-3)}$ **11.** $\frac{(2x+1)(2x-1)}{2x(x+1)^2}$

12. $\frac{x^2+15x+14}{10}$ **13.** $\frac{x-3}{x}$

14. $\frac{x(3x+4)}{4x^3+9x-36}$

15. $R_t=\frac{R_1R_2R_3R_4}{R_1R_2R_3+R_1R_2R_4+R_1R_3R_4+R_2R_3R_4}$

16.

$$R_t=\frac{R_1R_2R_3R_4R_5}{R_1R_2R_3R_4+R_1R_2R_3R_5+R_1R_2R_4R_5+R_1R_3R_4R_5+R_2R_3R_4R_5}$$

Reteaching with Practice

1. $\frac{x^2+6}{2x^2}$ **2.** $\frac{4x+9}{2x(x+3)}$ **3.** $\frac{7x+23}{(x+5)(x+1)}$

Lesson 9.5 *continued*

4. $\dfrac{x(x+4)}{(x+1)(x-1)}$ 5. $\dfrac{2(x-2)}{x}$

6. $\dfrac{-2x-14}{(x+2)(x-2)}$ 7. $\dfrac{3x^2+2}{x^3}$

8. $\dfrac{x^2-4x-18}{(x+3)(x+2)}$ 9. $\dfrac{x}{3(x-1)}$ 10. $\dfrac{2x-1}{x^2}$

11. $\dfrac{x+1}{x-1}$

Cooperative Learning Activity

Instructions

2. \$1444.13 **3. a.** \$391.70 **b.** \$415.18 **c.** \$451.90

Analyzing the Results

1. a. 1601.20 at 8% **b.** 2446.48 at 12% **c.** 3768.40 at 18%

Real-Life Application

1. 63,343 **2.** 1995

3. $\dfrac{\text{Number of branches}}{\text{Number of banks}} = \dfrac{45{,}020.26t^4 + 3{,}321{,}403.8t^2 + 50{,}543{,}000}{4683.52t^4 + 523{,}113.44t^2 + 12{,}282{,}300}$

4. about 6

Challenge: Skills and Applications

1. a. $f(x) = x + \frac{1}{x}$; $x = 2, -2$

b. 1; no, as $x \to 0$, $x + \frac{1}{x} \to +\infty$. **c.** $\sqrt[3]{2}$

2. -1 **3.** $\dfrac{1}{x-a}$ **4.** $\dfrac{x-b}{xb}$ **5.** $\dfrac{x+b}{b}$

6. $\dfrac{3}{x^2-2x}$ **7.** $\dfrac{-1}{a^2x^2}$

8. a. $A = \dfrac{1}{p-q}$, $B = \dfrac{1}{q-p}$

b. $\dfrac{1/5}{x-3} - \dfrac{1/5}{x+2}$

Lesson 9.6

Warm-Up Exercises

1. $-2, 1$ **2.** -3 **3.** -4 **4.** $-5, 5$ **5.** $0, -5, 5$

Daily Homework Quiz

1. $2(x^2-9)$ **2.** $6x^2(x+3)$ **3.** $\dfrac{x+5}{x-4}$

4. $\dfrac{10-3x}{4x^2}$ **5.** $\dfrac{2(3x-7)}{(x+2)(x-3)}$ **6.** $\dfrac{2x(x-3)}{3(x+12)}$

Lesson Opener

Allow 15 minutes.

1. $-7.62, 2.62$ **2.** 0.25, 15.75 **3.** 5.5 **4.** $-5, 3$ **5.** $-8, 0$ **6.** 0.75, 3.82 **7.** -1.25 **8.** $-3.21, -2.30, 1.17, 2.33$

Practice A

1. no **2.** yes **3.** no **4.** no **5.** $-\frac{1}{3}$ **6.** no solution **7.** $\frac{9}{7}$ **8.** -4 **9.** $-\frac{12}{7}$ **10.** 4 **11.** no solution **12.** $-5, -1$ **13.** 6 **14.** 2, 5 **15.** $-7, 4$ **16.** $-5, 6$ **17.** $\frac{9}{11}$ **18.** $-3, \frac{4}{3}$ **19.** 3 **20.** 3 **21.** no solution **22.** no solution **23.** 4.5 miles, 8 miles

Practice B

1. yes **2.** no **3.** no solution **4.** $-\frac{1}{3}$ **5.** 0 **6.** no solution **7.** -3 **8.** $\frac{7}{5}$ **9.** -5 **10.** 11 **11.** $-7, 4$ **12.** 3 **13.** $-4, 4$ **14.** -5 **15.** no solution **16.** $0, -2$ **17.** $5, -2$ **18.** $-2, 2$ **19.** $-\frac{1}{6}$ **20.** -1 **21.** 12,000 dozen cards **22.** 30 miles per hour

Practice C

1. 20 **2.** no solution **3.** 10 **4.** -2 **5.** -2 **6.** $-1, 3$ **7.** no solution **8.** $\frac{3}{17}$ **9.** 1, 3 **10.** -8 **11.** 2, 5 **12.** $\frac{7}{3}$ **13.** -2 **14.** no solution **15.** $-\frac{1}{2}$ **16.** $\frac{1}{2}$ **17.** $\frac{8}{5}$ **18.** 4 **19.** 1.17 (Jan.), 12 (Dec.) **20.** 50,000 baskets

Reteaching with Practice

1. $-\frac{1}{3}$ **2.** $-\frac{16}{3}$ **3.** -4 **4.** $\frac{1}{2}, -2$ **5.** $-1, 3$ **6.** $-3, 8$

Interdisciplinary Application

1. 1985 and 1992 **2.** 1994 through 1996

Lesson 9.6 *continued*

3. 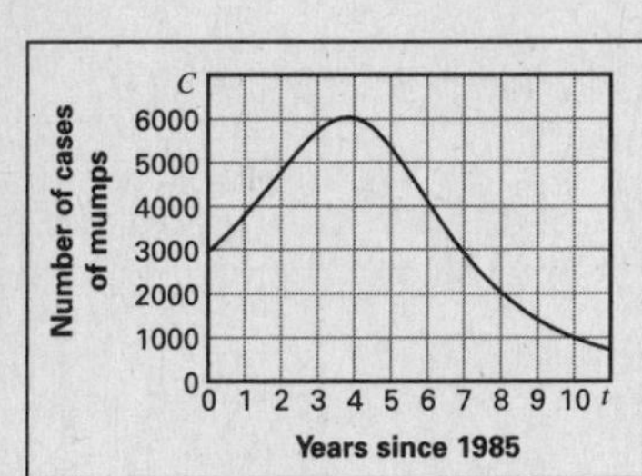

4. 1989

Math and History Application

1. about 0.5%
2. 20%, which agrees closely with the actual percentage of oxygen in the air at sea level
3. 39 atmospheres **4.** 1104 atmospheres
5. The pressure at the top of Everest is about 0.3 atmospheres.

Challenge: Skills and Application

1. -1 **2.** $-1, -2$ **3.** $1, \frac{3}{2}$ **4.** $4, 6$

5. $-4, -9$ **6.** $1, \frac{7}{2}$

7. The equation reduces to $6(x - 3) - x + x^2 = x(x - 1) \Longrightarrow 6x - 18 = 0$; $x = 3$. But this apparent solution makes the right side undefined.

8. about 5.5 mi/h

9. a. $\frac{d}{20} + \frac{d}{r}$ **b.** $\frac{2d}{\frac{d}{20} + \frac{d}{r}}$

c. $\frac{2d}{\frac{d}{20} + \frac{d}{r}} = 40$; no solution; the equation reduces to $0 = 20$.

Quiz 2

1. x^3y^2 **2.** $\frac{x + 2}{3x}$ **3.** $\frac{3x^2 + 14x}{(x - 4)(x + 3)}$

4. $\frac{7x^2 - 3x + 5}{(3x + 5)(3x - 5)}$ **5.** $\frac{33x + 21}{-6x + 2}$

6. $\frac{(5x - 1)^2(5x + 1)}{x}$ **7.** $\frac{-12x + 54}{x + 3}$

8. $x^2 + 4x - 6$ **9.** 1 **10.** $\frac{5}{2}, 3$

Review and Assessment

Chapter Review Games and Activities

1. $k = 12$ **2.** $y = 6$ **3.** $k = 3$ **4.** $z = 9$

5. $\frac{x + 2}{2x - 1}$ **6.** $\frac{x - 2}{x + 3}$ **7.** $6x$ **8.** $x = -63$

9. $x = 4, -1$ With a SPUDOMETER

Test A

1. $y = \frac{2}{x}$; 1 **2.** $y = -\frac{4}{x}$; -2 **3.** $y = \frac{12}{x}$; 6

4. $z = \frac{1}{8}xy$; 1 **5.** $z = -xy$; -8

6. 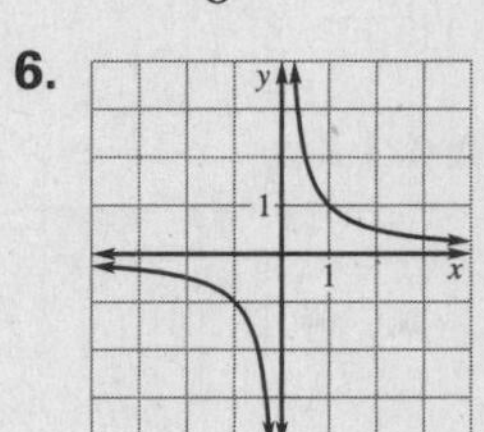

7. 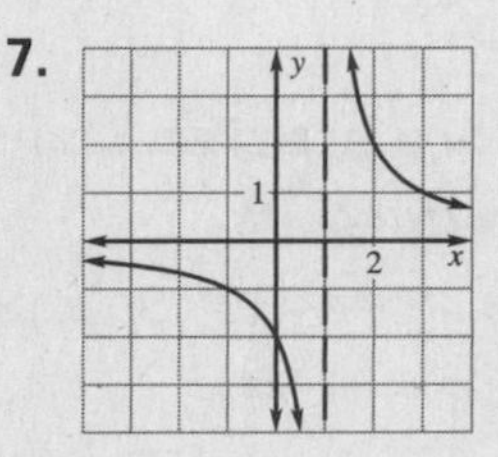

8. 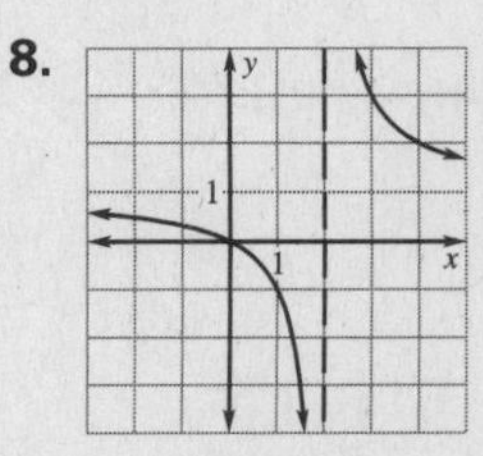

9. 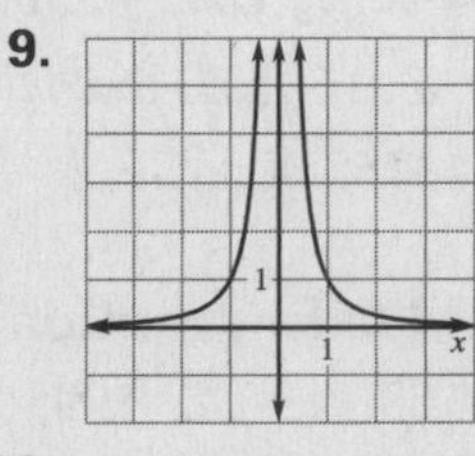

10. 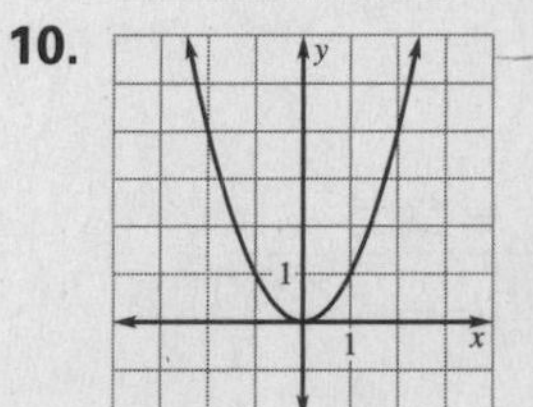

11. 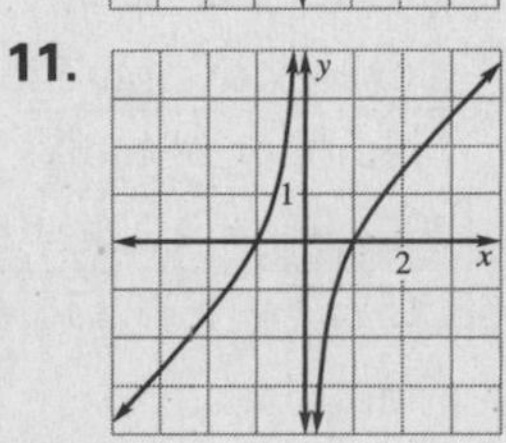

12. $\frac{x}{2}$ **13.** $\frac{x^2}{x + 1}$ **14.** 2 **15.** $\frac{5x}{x - 2}$

16. $\frac{-4x}{x - 3}$ **17.** $-\frac{3}{4x}$ **18.** $\frac{63}{32}$ **19.** $\frac{x(x - 12)}{3(5x + 1)}$

20. $\frac{(x + 3)^3}{18x^4}$ **21.** 2 **22.** $\frac{3}{4}$

23. -4 (7 is extraneous) **24.** $z = \frac{xy}{x + 2}$

25. 14,000 dozens of golf balls

Test B

1. $y = -\frac{4}{x}$; -1 **2.** $y = \frac{4}{x}$; 1 **3.** $y = -\frac{8}{x}$; -2

4. $z = -\frac{1}{4}xy$; 1 **5.** $z = -\frac{1}{2}xy$; 2

Review and Assessment *continued*

6.

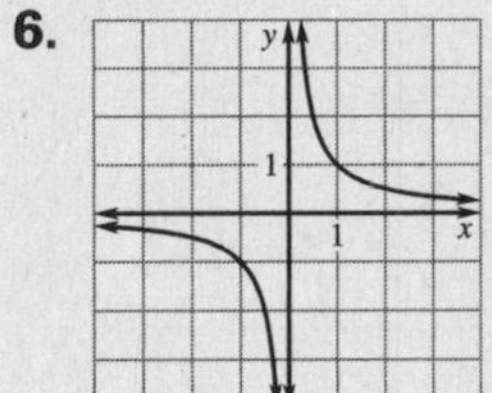

7.

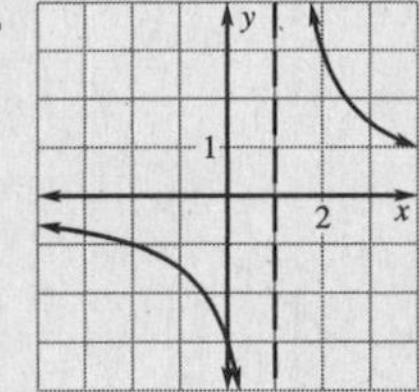

8.

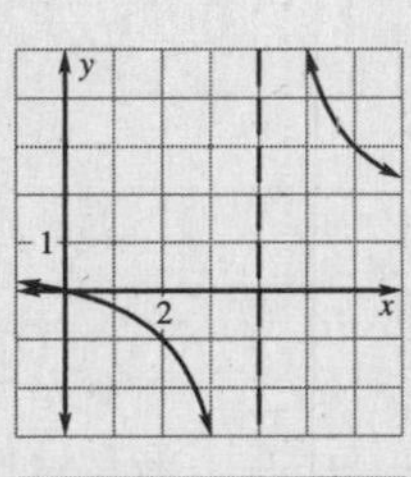

9.

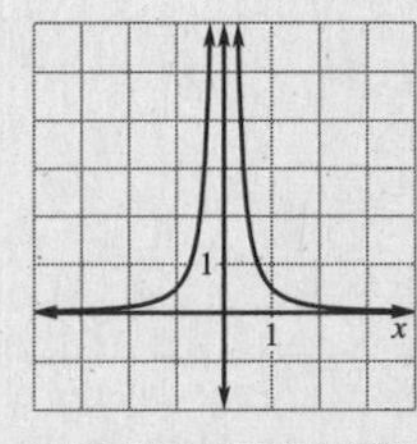

10.

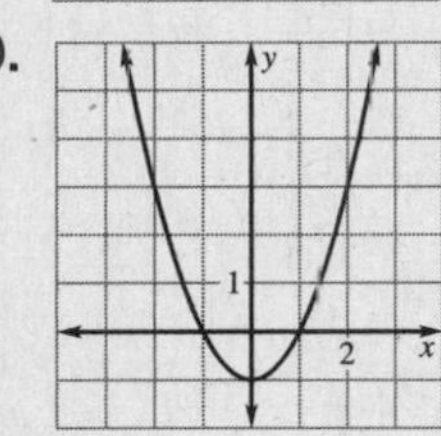

11.

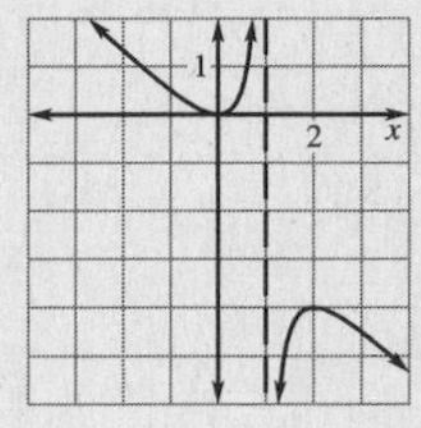

12. $\dfrac{11x}{2y}$ **13.** 1 **14.** $2x + 2$ **15.** $\dfrac{x-1}{x^2}$

16. 2 **17.** $2x$ **18.** $\dfrac{25}{x^2}$ **19.** $\dfrac{x-y}{x+y}$

20. $\dfrac{10xy + 15x}{15y + x^2y}$ **21.** 14

22. -1; (1 is extraneous) **23.** $\dfrac{3}{2}, \dfrac{7}{4}$

24. $z = \dfrac{y(x+4)}{x}$ **25.** 160,000 hats

Test C

1. $y = -\dfrac{9}{x}; -3$ **2.** $y = \dfrac{1}{x}; \dfrac{1}{3}$ **3.** $y = -\dfrac{2}{x}; -\dfrac{2}{3}$

4. $z = -\dfrac{1}{8}xy; -\frac{3}{4}$ **5.** $z = 9xy$; 54

6.

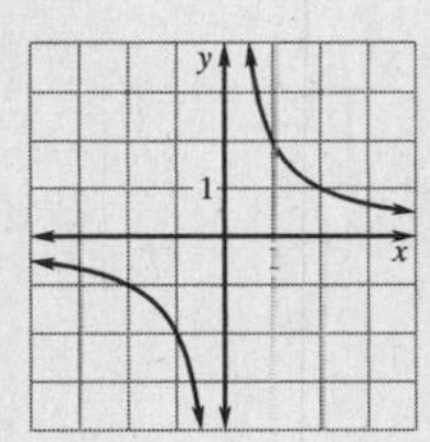

7.

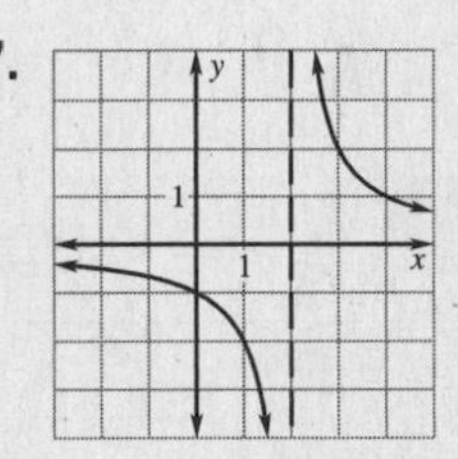

8.

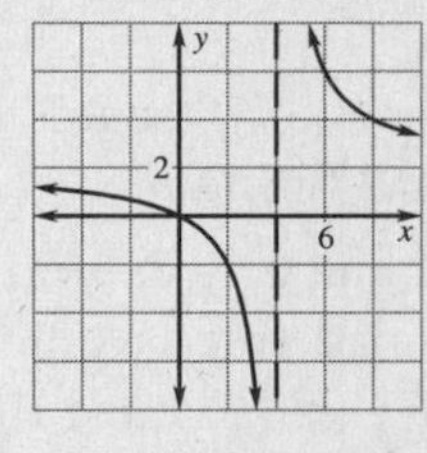

9.

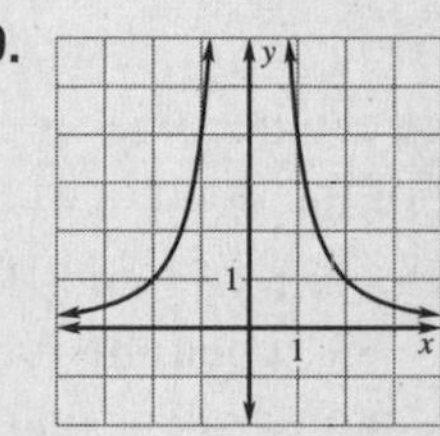

10.

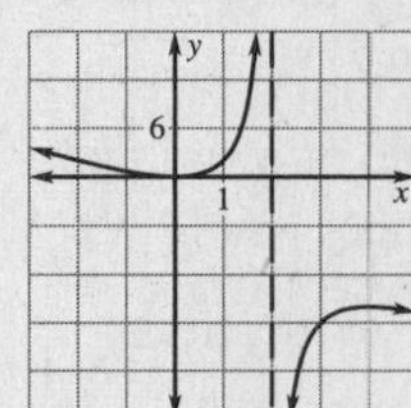

11.

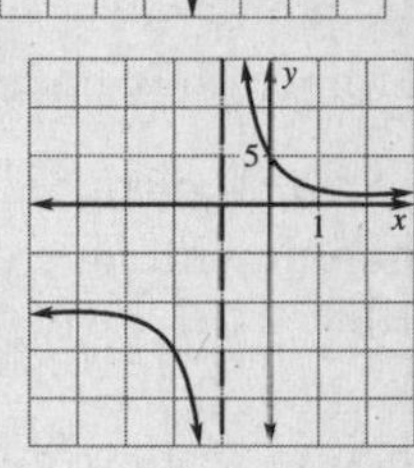

12. $\dfrac{5yz + 3xz - 2xy}{xyz}$ **13.** $\dfrac{3}{x+3}$ **14.** $\dfrac{7}{10}$

15. $8x(x+3)(x-2)$ **16.** $\dfrac{3(x+1)}{2x-3}$ **17.** -1

18. $\dfrac{2y^2 - 20x^2y}{7xy^2 + 15x^2}$ **19.** $\dfrac{20}{x+7}$ **20.** $\dfrac{x}{x-1}$

21. 4, -4 **22.** 5 **23.** 3, -2 **24.** $z = \dfrac{y(x-3)}{x}$

25. about 130 boxes

SAT/ACT Chapter Test

1. B **2.** A **3.** D **4.** C **5.** D **6.** A **7.** D **8.** C **9.** B

Alternative Assessment

1. a–d. Complete answers should address these points: **a.** • Explain that a common denominator must be found by multiplying the denominators. **b.** • Explain that a common denominator in this problem would be the same as part **(a)**, except that only i would need to change. **c.** • Explain that in multiplication the numerators and denominators are multiplied and the result is simplified by canceling/reducing common factors. **d.** • Explain that when dividing, you must multiply the first fraction by the reciprocal of the second fraction. **2. a.** Equation I solution is $x = -\frac{14}{3}$; equation II solution is $x = 2$; an extraneous root occurred in equation II; multiplying through by $(x + 3)$ caused this extraneous root to occur. **b.** *Sample answer:* Extraneous roots can occur when multiplying both sides of an equation by an expression. The number that would make the expression zero cannot be multiplied through, making it an "illegal" operation for that value. This

Review and Assessment *continued*

value can become an extraneous root when it is not part of the domain of the original function.

c. Domain of I and II: all real numbers except $x = -3$, when the function is undefined, vertical asymptotes occur on the graph. **d.** In I and II the function is undefined when $x = -3$.

e. *Sample answer:* An example of a rational function with no extraneous roots would be $\frac{10}{x+3} + \frac{10}{3} = 6$. An example of a rational function with extraneous roots would be $\frac{x}{x+2} = \frac{8}{x^2-4} + 2$. The second function has an extraneous root at $x = -2$. This cannot work because it would make the denominator equal to zero. **3.** *Sample answer:* In solving radical equations, when both sides of the equation are squared, an extraneous root can occur. This is similar to rational equations when there are variables on both sides of the equation. The sides are multiplied by themselves, creating a multi-solution equation. Graphing can help to show where the two graphs intersect. It also illustrates any undefined values.

Project: The Ideal Container

1. $r \approx 4.30$ cm, $h \approx 8.60$ cm; $S \approx 349$ cm^2

2. About 7.94 cm by 7.94 cm by 7.94 cm; $S \approx 378$ cm^2 **3.** About 5.72 cm by 11.45 cm by 7.63 cm; $S \approx 393$ cm^2 **4.** *Sample answer:* The most efficient shape is a sphere of radius 4.924 cm; its surface area is 305 cm^2. Most other shapes are less efficient than the cylinder.

5. *Sample answer:* The cylinder is not the most efficient. Manufacturers use the cylinder because it is reasonably efficient, it is easy to manufacture and store, and it is easy for consumers to open using a can opener.

6. $r \approx 5.42$ cm; $h \approx 5.42$ cm **7.** *Sample answer:* Other factors include durability, safety, and consumer preferences.

Cumulative Review

Chapter 9

1. 3 or $-\frac{7}{3}$ **2.** $-\frac{7}{2}$ or $-\frac{17}{2}$ **3.** 56 or -8

4. $n > 2$ or $n < -1$ **5.** $8 \geq n \geq -4$

6. $x \geq 7$ or $x \leq -29$

7.

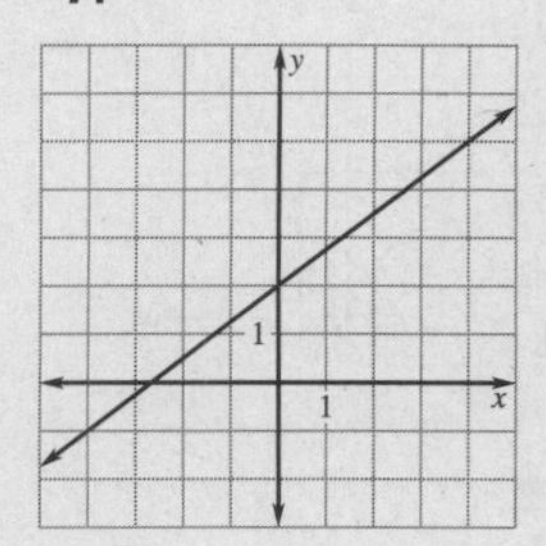

8.

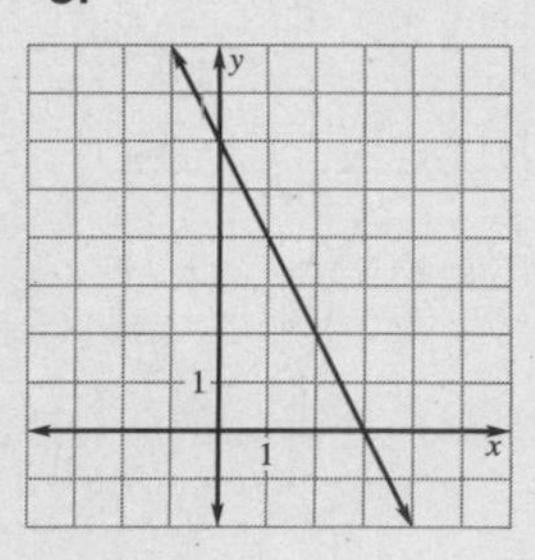

9.

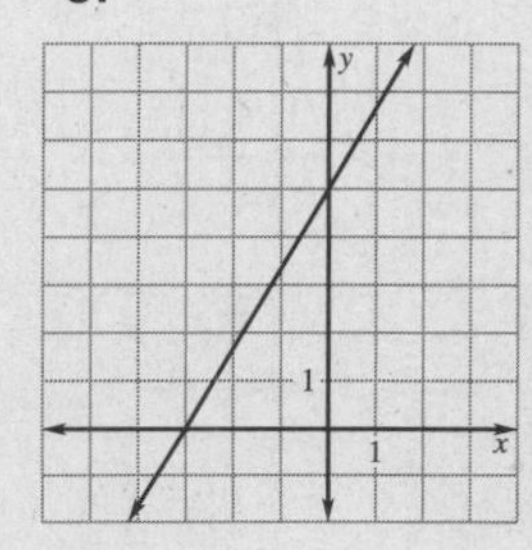

10.

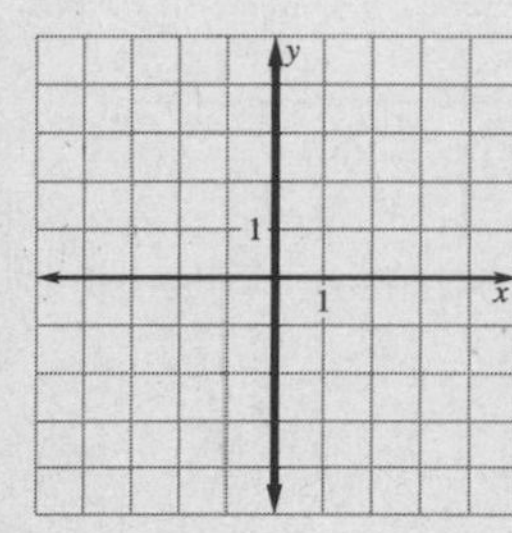

11.

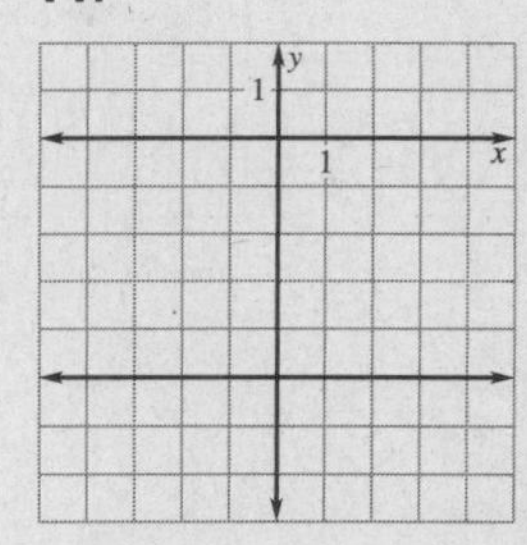

12.

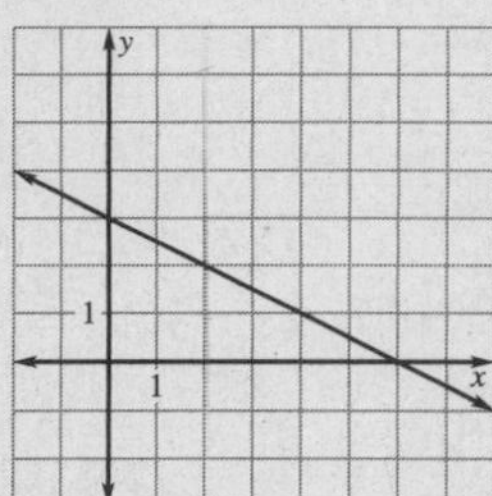

13. C **14.** A **15.** B **16.** infinite **17.** none

18. one

19.

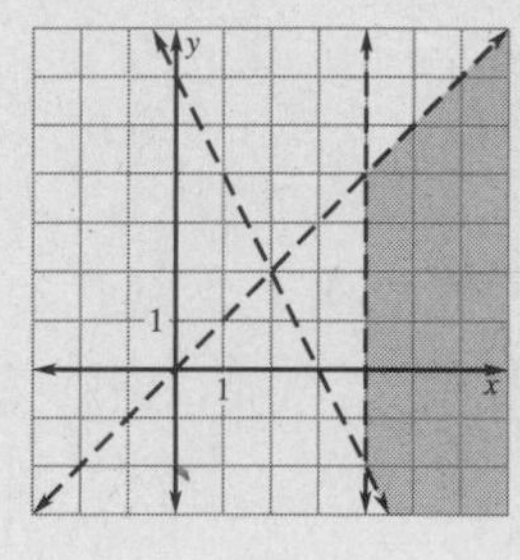

20.

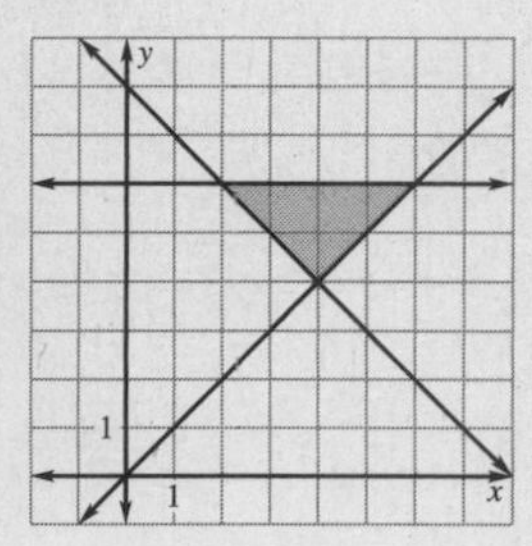

21.

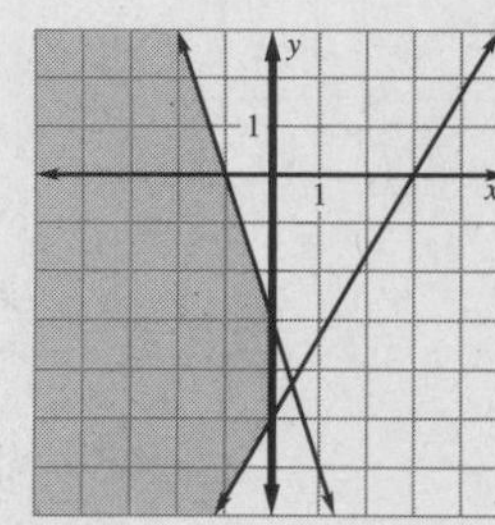

22. $\begin{bmatrix} 14 & -7 \\ 8 & 12 \\ 2 & -16 \end{bmatrix}$

Review and Assessment *continued*

23. $\begin{bmatrix} 9 & -15 \\ 13 & -3 \\ 5 & 15 \end{bmatrix}$ **24.** $(x + 1)(x + 6)$

25. $(3x - 5)(3x - 10)$ **26.** $3(3x - 1)(x - 1)$

27. $(7x + 9)(7x - 9)$ **28.** $(3y - 8)(3y - 2)$

29. $x(x + 32)$ **30.** $\sqrt{34}$ **31.** 3 **32.** $\sqrt{257}$

33. $\sqrt{5}$ **34.** $3\sqrt{2}$ **35.** $\sqrt{17}$

36. $-2 \pm i\sqrt{14}$ **37.** $\dfrac{3 \pm i\sqrt{7}}{2}$ **38.** $3, -1$

39. $\dfrac{-5 \pm \sqrt{37}}{2}$ **40.** $\dfrac{2 \pm \sqrt{10}}{2}$ **41.** 4, 2

42. $y = 3(x + 4)^2 + 6$

43. $y = -2(x - 1)^2 - 9$ **44.** $y = -x^2 + 3$

45. x^8y^3 **46.** $\dfrac{y^5}{3x}$ **47.** $\dfrac{y^6}{16x^4}$ **48.** x^3y^{14}

49. $\dfrac{-3}{x^2}$ **50.** x^6 **51.** 17 **52.** 281 **53.** 18

54. -25 **55.** $f(x) = -6x + 13$; first; linear; -6

56. $f(x) = \frac{1}{3}x^3 + 2x + 8$; third; cubic; $\frac{1}{3}$

57. no **58.** $2x^2 + 6x - 9$; all real numbers

59. $6x + 9$; all real numbers

60. $2x^2 + 12x$; all real numbers

61. $x^4 + 6x^3 - 9x^2 - 54x$; all real numbers

62. $x^4 - 12x^2 + 27$; all real numbers

63. $x^4 - 18x^2 + 72$; all real numbers

64. 8 units left **65.** 6 units left, reflected over x-axis **66.** 1 unit left, 5 units up **67.** 1 unit right

68. $1; y = 0$ **69.** $-3; y = 0$ **70.** $4; y = 3$

71. $\frac{1}{2}; y = 0$ **72.** $\frac{3}{2}; y = 0$ **73.** $10; y = 5$

74. $\dfrac{e^{2x}}{3}$ **75.** $3e^{3x}$ **76.** $3e^{3x+1}$ **77.** 2.398

78. 0.239 **79.** 1.375